Lecture Notes in Energy

Volume 21

For further volumes:
http://www.springer.com/series/8874

L.M. Abadie · J.M. Chamorro

Investment in Energy Assets Under Uncertainty

Numerical methods in theory and practice

Springer

L.M. Abadie
Basque Centre for Climate Change (BC3)
Bilbao
Spain

J.M. Chamorro
Department of Financial Economics II
University of the Basque Country
UPV/EHU
Bilbao
Spain

ISSN 2195-1284 ISSN 2195-1292 (electronic)
ISBN 978-1-4471-5591-1 ISBN 978-1-4471-5592-8 (eBook)
DOI 10.1007/978-1-4471-5592-8
Springer London Heidelberg New York Dordrecht

Library of Congress Control Number: 2013949432

Printed on acid-free paper

Springer is part of Springer Science+Business Media (www.springer.com)

To Esperanza

To Inmaculada

Preface

Valuation is intrinsic to anyone confronted with a choice. One may have to choose between a ticket to the cinema and a frugal supper. Another one must suddenly decide whether to jump in tune with the herd or stay quiet instead. As far as humans are concerned, clearly we take account of the world around us, and we are well aware of time and uncertainty. But there are other inputs to the valuation process. Prominent among them are how we trade off present consumption against future consumption, our (faulty) prowess at assessing probabilities, and our (changing) appetite for risk. In sum, valuation involves both objective data and human nature.

On the other hand, energy has become a fundamental pillar of the world's increasing prosperity and economic growth. We are more or less used to taking affordable and reliable energy for granted. But the list of required aspects is getting longer, with clean energy and sustainable energy use now firmly ingrained. Leaving aside the potential tensions among these four goals, two key features that compound the whole issue are the large capital investments involved and the long turnover times of energy systems. These features in turn push managers toward thinking twice (so to speak) when they assess investments in energy assets.

To the extent that the energy sector is part of the whole picture, we deal with investments that simultaneously display the following characteristics: the return on these investments is uncertain (sometimes, even the return *of* them is), they are irreversible to a high degree, and managers have a number of flexibility options at hand (e.g., the option to delay investment, or to alter the scale of the project, to temporarily cease operation, to abandon it completely, ...). These features cannot be properly accounted for by traditional discounted cash flows valuation methods. The Real Options (RO) approach, however, is better equipped to cope with them.

Industry practitioners will thus (hopefully) find this book useful. They will get better estimates of the value of an energy project and of the option to invest in it. This is important not only to project developers, but to the funding institutions as well. Indeed, it is important for the broader community. For one, the financial crisis has spelt ill for a number of energy projects. Thus, even profitable projects at first sight can have a hard time in getting funded. At the same time, public finances right now are harder pressed than usual. Support schemes, in particular, need consequently to be re-assessed, to target the appropriate projects and do so in the right doses. Indeed, government policies are not only needed to support invention and stimulate innovation. They are also needed to align the market forces and accelerate the speed of adoption of the most promising cleaner energy technologies, so that they progress down the learning curve toward market competitiveness. In doing so, they will become widely deployed and reach commercial scale. This acceleration is needed because the world can hardly afford to wait for long-term solutions; the pressure on the system keeps on mounting (for instance, think of climate change), and it is already daunting.

With this background, we have strived to be as clear as possible when introducing the RO valuation approach to a relatively broad audience. This has led us to draw on a number of handbooks and articles, those which we felt were more accessible and friendly to the non-expert. To ease the learning process, we include a number of examples which are solved under the traditional "paper and pencil" perspective. Of course, more complex settings require more advanced tools, which are also explained in the text.

The book is organized into three parts. Part I comprises a single chapter, which restricts itself to a context without risk. Chapter 1 discusses the basics of compounding and discounting over time. Then it addresses the valuation of finite-lived annuities and also perpetual ones. This part also introduces dynamic programming as an optimization technique that is frequently adopted in the subsequent chapters. It stresses the point that maximizing the value of an asset typically calls for the optimal management of that asset.

Part II moves to an environment characterized by both time and uncertainty. Chapter 2 explains the portfolio selection problem faced by a single investor in a mean-variance context. If the aggregate of investors is assumed to behave this way, then it is possible to come up with a well-known pricing method, namely the Capital Asset Pricing Model. Nonetheless, this is not the only pricing method. This chapter explains the so-called risk-neutral valuation; it also introduces the basics of futures markets. As will be clear in the coming chapters, the former can be used consistently with data from the latter. Next, Chap. 3 shows two stochastic processes that have been widely used to characterize the price behavior of a number of assets. We explore in depth their properties both in the physical world and the risk-neutral world. From the behavior of these (underlying) assets, we then address the valuation of derivative assets on them, like annuities, futures contracts, and (investment) options. The next three chapters involve numerical methods, which are frequently needed because of the lack of analytical solutions to the valuation models. Chapter 4 develops a number of binomial lattices. Chapter 5, instead, shows several finite difference methods. Chapter 6 shows how to run Monte Carlo simulation. There are several examples solved with the three methods, so their results can be easily compared to each other.

Last, Part III comprises five chapters with a strong focus on investments in energy assets. Chapter 7 collects some real data about power generation technologies. By blending technical parameters and market prices it is possible to develop some metrics of profitability. The latter depends on a number of factors, among them the thermal efficiency of the plant, the availability rate, whether it operates under carbon constraints, and so on. Next we show how to use both the analytical solutions and the numerical methods by demonstration. Chapter 8 considers just one source of risk, say, natural gas price. Chapter 9 accounts for two risk factors, e.g., coal price and electricity price. And Chap. 10 supports up to three sources of risk, say the above two plus carbon allowance price. Lastly, in Chap. 11 we extend the valuation process to a case with active management of a power plant that can be switched "on" or "off."

The book itself is somewhat of a matter of chance. Were it not for our common friend, José M[a] Pérez de Villarreal, our paths would have never crossed almost surely. We met for the first time in early 2004. One of us (Luis) worked at a financial institution, while the other (José) was engaged in academia. Luis set out to develop a Ph.D. dissertation on energy economics, with José in the role of advisor. By then, José was familiar with RO and had contacted several professors in this area, among them Arturo Rodríguez, Fernando Gómez-Bezares, Juan Mascareñas, and Prosper Lamothe. The endeavor was successful, and in late 2007 the Ph.D. candidate got the approval of the Thesis Committee: Margaret Armstrong, Inmaculada Gallastegui, Gonzalo Cortázar, Juan Pablo Montero, and Richard de Neufville.

Since then we have published several papers in academic journals and presented them at national and international conferences, where we have met a number of researchers, learnt from their expertise, and benefitted from their feedback. The Annual Conference on Real Options allowed us to get in contact with Lenos Trigeorgis, Gordon Sick, Dean Paxson, Marco Antonio Díaz, Luiz Brandão, Susana Alonso, and Gabriel de la Fuente, among others. Similarly, the Annual Congress of the Spanish Association for Energy Economics gave us the opportunity to meet Emilio Cerdá, Francisco J. André, Pablo Arocena, Antonio Canoyra, Dolores Furió, Xabier Lavandeira, Pedro Linares, Enrique Loredo, Pablo del Río, and Gonzalo Sáenz de Miera, to name a few. In the meantime, Luis joined the Basque Centre for Climate Change (BC3), where he develops his agenda as a Research Professor.

We gratefully acknowledge financial support over these years from the Spanish Ministry of Science and Innovation (ECO2011-25064), the Basque Government (IT-799-13), the University of the Basque Country UPV/EHU, and Fundación Repsol through the Low Carbon Programme joint initiative.[1] Luis also thanks colleagues at the Basque Centre for Climate Change (BC3) Ramón Arigoni, Dirk Rubbelke, and Nerea Ortíz. José shows his gratitude to fellows at the University of the Basque Country UPV/EHU Juan Félix Jauregui-Arraburu, Txomin Iturralde, and Miguel Ángel Pérez. Last, but by no means least, we express our deepest gratitude to our colleagues and friends at the Low Carbon Programme initiative, Mari Carmen Gallastegui, Marta Escapa, Anil Markandya, Alberto Ansuategui, Ibon Galarraga, and Mikel González-Eguino, for their trust and continued support.

[1] http://www.lowcarbonprogramme.org

Contents

Part I
Investment Under Certainty

A journey of a thousand miles begins with a single step. Lao-tzu (604 BC–531 BC).

Chapter 1
Valuation Made Simple: No Uncertainties, Just Time

1.1 Some Preliminaries

Energy assets like oil wells or power plants typically have decades-long useful lives. Besides, the pattern of costs and revenues over these lives is far from symmetric. A sizeable chunk of total costs appears in the early stages in the form of upfront sunk costs while revenues only start to accrue once the project is up and running. For example, natural gas-fired stations take less time to build than nuclear plants. If the construction period coincides with huge outlays then it can be necessary to tap the capital market for getting funds (these will be later amortized with incoming revenues). Time is thus is a major driver in the valuation of energy assets.

Needless to say, a dollar today and a dollar 1 year from now are not worth the same. There are several reasons for this. For instance, the economic life cycle of many consumers/savers usually starts in the red (so to speak); by the time it turns to black they are well into their forties or even fifties. A changing price index can also explain the difference. Thus, if prices are expected to rise on average then the purchasing power of a dollar note will fall. These examples are just mere manifestations of the *time value of money* in finance jargon. See Fisher (1907, 1930).

The above arguments lead us naturally to introduce some preliminary material regarding interest, interest rates, and compounding. Yet there is one more reason for paying attention to it. Specifically, this book makes extensive use of the information content embedded in market prices. A number of commodities are regularly traded on forward and futures markets, among them oil, coal, natural gas, electricity, and emission allowances. Once we are familiar with the basics of interest rate theory we can safely move on to determining forward and futures prices.

1.1.1 Simple and Compound Interest

Let A denote a certain amount of money. Assume that this sum is invested in a bank deposit bearing an interest rate r per annum over a whole year. The future value FV of this account at the end of the year is:

L.M. Abadie and J.M. Chamorro, *Investment in Energy Assets Under Uncertainty*, Lecture Notes in Energy 21, DOI: 10.1007/978-1-4471-5592-8_1,

$$FV = A(1 + r). \tag{1.1}$$

If, instead, the sum is placed at the bank for n years then we have two possibilities (depending on the particular clauses agreed with the bank in the contract):

$$FV = A(1 + rn), \tag{1.2}$$

$$FV = A(1 + r)^n. \tag{1.3}$$

Equation (1.2) provides the total value after n years under the simple interest rule. In this case the balance of the account grows linearly with time n (measured in years). Now consider the case in which the total value at the end of the first year, $A(1 + r)$, becomes the base for the second year; this is the essence of compounding. Then, at the end of the second year FV will amount to $A(1 + r)^2$. After n years of annual compounding the terminal value is given by Eq. (1.3).

When compounding takes place more frequently, say, m times per year, the (annual) interest rate r over the period involved applies accordingly. Thus, with quarterly compounding ($m = 4$) after 1 year FV will be:

$$FV = A(1 + \frac{r}{4})^4. \tag{1.4}$$

By the same token, in general after n years we would have:

$$FV = A(1 + \frac{r}{m})^{mn}. \tag{1.5}$$

Continuous compounding Now we are ready to move from discrete compounding, as in Eqs. (1.4) and (1.5), to continuous compounding. We just need to think of arbitrarily short periods of time between successive rounds of compounding along the year. In other words, the rate of compounding becomes arbitrarily high ($\boldsymbol{m} \to \infty$) so the (instantaneous) interest rate (r/m) tends to zero while the power (m) grows without bound. We can rewrite Eq. (1.5) in a different way:

$$FV = A(1 + \frac{r}{m})^{mn} = A\left[(1 + \frac{r}{m})^{\frac{m}{r}}\right]^{rn} = A\left[(1 + \frac{1}{g})^g\right]^{rn},$$

where we have defined $\boldsymbol{g} \equiv \boldsymbol{m}/\boldsymbol{r}$. As $\boldsymbol{m} \to \infty$ also $\boldsymbol{g} \to \infty$. By now many readers will be guessing that the number $\boldsymbol{e} = 2.7182\ldots$ (the base of the natural logarithms) is at work here since:

$$\lim_{g \to \infty} (1 + \frac{1}{g})^g = e.$$

In the end we get:

$$FV = Ae^{rn}. \tag{1.6}$$

Thus, if we initially invest an amount $A = \$100$ continuously compounded at a rate of nine percent ($r = 0.09$), after 1 year ($n = 1$) we would receive $FV = 100\, e^{0.09} = 109.4$. Hence we can easily check that under continuous compounding the growth rate would be $e^{rn} = 1.0941$ in gross terms, or $1.0941 - 1 = 0.0941$ in net terms; this effective rate is clearly higher than the nominal rate. For an arbitrary length of time t (in years) the formula for the terminal value is:

$$FV = Ae^{rt}. \tag{1.7}$$

Unlike what happens under the simple interest rule, with (continuous) compounding the balance of the bank account grows exponentially with time.

Given the simple interest rate it is easy to derive its equivalent rate in continuous time. If they both are to be really equivalent then we have $1 + r_s = e^r$. For example, if the (1 year) simple rate is $r_s = 0.09$, the continuous rate will be $\ln(1 + r_s) = r = 0.0862$. Indeed, for $t = 1$ simple interest yields $FV = A(1 + 0.09) = 1.09A$, while continuous compounding implies $FV = Ae^{0.0862} = 1.09A$.

1.1.2 Discounting

Once we have developed the formulae for the future value FV of a current deposit we can also undertake the opposite calculation, namely to derive the present value PV of a certain, future amount of money. Restricting ourselves to compound interest, it is clear from Eq. (1.3) that a dollar 1 year from now ($V = 1$, $n = 1$) is worth $\mathbf{A = 1/(1 + 0.09) = 0.9174}$ dollars as seen from today if the prevailing interest rate is 9 %. As before, continuous compounding reinforces the effect. Thus, Eq. (1.8) shows that the present value of \$1 in a year's time is a bit lower, namely $\boldsymbol{e^{-r} = 0.9139}$ if 9 % is the rate of continuous compounding (but just the same 0.9174 with the equivalent rate 0.0862). More in general, discounting a future amount of money V at a continuously compounded rate r over a time interval of length t yields a present value:

$$PV = A = Ve^{-rt}. \tag{1.8}$$

1.2 Cash Flow Streams: Annuities, and Perpetuities

Once we feel confident about computing the present and future values of a single amount of money we can proceed to the valuation of a series of cash flows occurring at pre-determined dates. Of course this is the case of a number of financial contracts, e.g. bonds or mortgages (remember that we neglect any consideration of uncertainty for the time being); see Luenberger (2009). But it is easy to think of similar examples also in industry. For instance, a given investment at an industrial facility to enhance energy efficiency may bring about a saving (assumed known for certain) in raw materials (say, coal) over the facility's life (also assumed deterministic). Or a technical device placed at the end of the pipe or the smoke-stack can allow a firm to abate emissions by some amount with the ensuing savings in terms of allowances required until decommissioning.

An important remark is in order. Uncertainty about the future cash flows from investment projects render them inherently risky. The *PV* of each cash flow should thus involve discounting at a rate adjusted for risk (this issue will be addressed in Chap. 2). When future cash flows are known for certain, however, they must be discounted at the riskless rate. This is just the case we explain next: the annuities are assumed to be risk free. Nonetheless, it is perfectly possible that the net balance of buyers and sellers of riskless assets changes across maturities. This means that the risk-free interest rate need not be constant over time (i.e. the term structure of interest rates need not be flat). For the sake of simplicity, though, we assume a flat rate r. Note also that this rate can also be used for discounting purposes when risky positions are hedged through contracts traded on futures markets.

Last, it is important to bear in mind that the key ingredients for valuation are cash flows (and not the accounting expenses or revenues, which can differ significantly from cash flows). Only accounting decisions with an impact on cash flows (such as the amortization schedule, which affects future tax payments) must be taken into account.

1.2.1 Annuities

As a starting point, consider the case of a known collection of cash flows $(A_0, A_1, A_2, \ldots, A_{n-1})$. We can think of them as regular deposits at a savings bank, yet nothing precludes them from taking on negative values (or zero). For simplicity assume that the first deposit A_0 is placed right now (at time $t = 0$), the next one A_1 is to be placed a year later ($t = 1$), then A_2 in 2 years' time ($t = 2$), and so on and so forth until the last cash flow A_{n-1} is deposited at $t = n - 1$. If the rate of compounding is annual ($m = 1$) then the *FV* of this stream (at $t = n$) is just a sum of future values:

$$FV = A_0(1+r)^n + A_1(1+r)^{n-1} + A_2(1+r)^{n-2} + \cdots + A_{n-1}(1+r). \quad (1.9)$$

This is because A_0 will stay for n years in the account; A_{n-1}, instead, will be placed and retrieved after just 1 year. Equation (1.9) can be written in a more compact way as:

$$FV = \sum_{k=0}^{k=n-1} A_k(1+r)^{n-k}. \quad (1.10)$$

Now, instead, consider the sequence of cash flows $(A_1, A_2, \ldots, A_n)$. Think of them as occurring at the end of each period. Thus, A_1 takes place at the end of the time elapsed between dates $t = 0$ and $t = 1$. Assume we want to compute the *PV* of the whole sequence. This comes down to computing the present value of each one of them and adding all them up:

$$PV = \frac{A_1}{(1+r)} + \frac{A_2}{(1+r)^2} + \cdots + \frac{A_{n-1}}{(1+r)^{n-1}} + \frac{A_n}{(1+r)^n}. \tag{1.11}$$

This sum can be expressed more succinctly as:

$$PV = \sum_{k=1}^{k=n} \frac{A_k}{(1+r)^k}. \tag{1.12}$$

A typical annuity is a financial contract whereby a pre-determined number of equal payments *A* take place periodically, starting at the end of year 1 and finishing at the end of period *n*. The formula for the *PV* of this common annuity is:

$$PV = \sum_{k=1}^{k=n} \frac{A}{(1+r)^k} = A \sum_{k=1}^{k=n} \frac{1}{(1+r)^k}. \tag{1.13}$$

The right hand of Eq. (1.13) shows the sum of a finite number of terms of a geometric progression. This can be proven to be:

$$PV = \frac{A}{r}\left[1 - \frac{1}{(1+r)^n}\right]. \tag{1.14}$$

In Sect. 1.2.4 below we prove a similar claim in a more general case; Eq. (1.14) happens to be a particular case (with flat annual payments).

1.2.2 Perpetual Annuities

In the specific case that the annuity makes a constant payment forever we are dealing with a perpetual annuity or perpetuity. In this case, the formula for the *PV* of this infinite sequence of payments simplifies to

$$PV = A \sum_{k=1}^{\infty} \frac{1}{(1+r)^n} = \frac{A}{r}. \tag{1.15}$$

The above formulas are relatively simple. In the chapters that follow we will derive more complex expressions that are valid when the payments change over time according to some stochastic process. This way we will get an estimate of the *PV* of saving one ton fuel (say, coal) per year over a number of years when its price is time varying. In principle this amount could be compared to the present cost of the investment required to accomplish that reduced consumption.

1.2.3 Annuities and Perpetuities Under Continuous Compounding

When an amount A is received from $t = 0$ on a continuous basis along the year over n years the PV is given by:

$$PV = A\int_0^n e^{-rt}dt = \frac{A}{r}\left[1 - e^{-rn}\right]. \tag{1.16}$$

Thus, for $A = 1{,}000$, $r = 0.0862$, and $n = 10$ we compute $PV = 6{,}701.66$.

Instead, in discrete time with $r = 0.09$, from Eq. (1.14) we would get $PV = 6{,}417.66$. Both rates have been shown above to be equivalent for discounting a cash flow so, why this gap? In the continuous case the annuity starts to be received earlier; consequently the PV is higher.

1.2.4 Increasing Annuities

Discrete time Now consider the case in which the annual payment grows by a constant factor g from 1 year to the next (starting from the initial level A_0). The PV of this annuity in discrete time is computed as:

$$PV = \frac{A_0}{(1+r)} + \frac{gA_0}{(1+r)^2} + \cdots + \frac{g^{n-1}A_0}{(1+r)^n}. \tag{1.17}$$

To derive this value we go through an intermediate stage, in particular:

$$\frac{(1+r)PV}{g} = \frac{A_0}{g} + \frac{A_0}{(1+r)} + \cdots + \frac{g^{n-2}A_0}{(1+r)^{n-1}} = \frac{A_0}{g} + PV - \frac{g^{n-1}A_0}{(1+r)^n}. \tag{1.18}$$

After a bit of algebra we come up with the final formula:

$$PV = \frac{A_0}{(1+r-g)}\left[1 - \frac{g^n}{(1+r)^n}\right]. \tag{1.19}$$

For $g = 1$ we deal with a constant annuity, and thus we are back to Eq. (1.14). For example, with $A = 1{,}000$, $r = 0.09$, $g = 1.05$, and $n = 10$ we get $PV = 7{,}798.43$.

In the case of an ever growing perpetual annuity the second term in brackets vanishes:

$$PV = \frac{A_0}{(1+r-g)}. \tag{1.20}$$

With $A = 1{,}000$, $r = 0.09$, and $g = 1.05$ we get $PV = 25{,}000$. This formula only makes sense provided $1 + r > g$.

Continuous time In continuous time, the payment grows at a compound rate α. The amount received at time t must be discounted at the compound rate r:

$$PV = A_0 \int_0^n e^{(\alpha - r)t} dt = \frac{A_0}{r - \alpha}\left[1 - e^{(\alpha - r)n}\right]. \tag{1.21}$$

If $r > \alpha$, as $n \to \infty$ Eq. (1.21) converges to:

$$PV = \frac{A_0}{r - \alpha}. \tag{1.22}$$

For $A = 1{,}000$, $r = 0.0862$, $\alpha = 0.05$, and $n = 10$ we compute $PV = 8{,}389.99$.

1.3 Management and Value

Up to now we have dealt with the valuation of cash flows considering them as exogenously given. In other words, the decision maker simply takes them for granted as if there were no way to influence their size or timing. In this case there is no room for management at all. The only decision comes down to the simplest choice: take it or leave it. This situation resembles pretty much a "buy-and-hold" strategy in the stock market: the whole point is to pick carefully the appropriate stocks at one time and then keep the portfolio unchanged until the end of the investment horizon. Such passive management may be optimal under some particular circumstances but arguably this is more the exception than the rule. The usual scenario is one in which the value of the investment project depends intimately on how the project is managed and even designed; see de Neufville and Scholtes (2011). Good management typically translates into superior performance of the project. Hence, if the value of the firm (its portfolio of projects) is to be maximized then optimal management (of the projects) becomes essential.

We can illustrate this general principle in a number of different contexts. They range from mutual fund managers to oil extracting firms to fishermen with clearly defined property rights over a fish stock. Intuitively, happening to receive major inflows just in the periods with higher returns translates into a better performance than another manager unlucky enough to annotate those inflows in the 'right' period. Similarly, extracting more of the stuff when prices are relatively high necessarily makes a difference with respect to doing so when the prices are low. Yet there is a crucial subtlety here: how we exploit the resource today not only affects today's profits but also tomorrow's since the stock left at the end of period 0 is just the stock we start from in period 1. In sum, optimal management cannot be addressed on a period-by-period basis as if these periods were independent from each another. Since present decisions impact future results the whole sequence of optimal decisions at any time must be solved at once. Fortunately mathematicians provide us with the suitable tool to accomplish this feat (Bellman 1957).

1.4 Dynamic Programming

1.4.1 A Friendly Introduction: Charting the Shortest Route

A usual example to explain optimal decision making in a dynamic context is that of a salesman with the firm's management concerned about travel expenses (we can similarly think of a tank wagon full of precious gasoline); see Cerdá (2001). The context is one in which there are a number of possible routes to get from city A (say, Albany NY) to city H (say, Houston TX) in four days. Some other cities (B, C, D, E, F, G) are scattered along the way; thus one possible route is A–C–E–F–H. Since each route entails a different length, the problem consists in finding the shortest way from A to H crossing some of the cities in between.

It is useful to build a graphic representation of this problem. For example, each city can be represented by means of a node suitably labeled with the corresponding letter. An arrow starting from a node and reaching another one stands for the road between the cities involved. A number attached to each arrow shows the distance from the city of departure to that of arrival (Fig. 1.1).

Of course this representation is most useful when the scale of the problem is relatively limited as in the above example. Indeed, one can identify the shortest route by brute force (see Table 1.1). A single table with the whole collection of alternative paths alongside the distances involved in each case will suffice to determine that the best course is A–B–E–F–H. Anyway we can resort to another procedure which unlike the former is not direct but is quite convenient when the dimension of the problem increases; its name: dynamic programming.

Needless to say, it is crucial to frame the map appropriately: starting from A, we have as many arrows as intermediate cities at a one-day driving distance, say B and C; similarly for the second and third days. Thus, cities D and E can be within reach from either B or C after spending a whole day on the road. In this case we have two arrows connecting B with D and E; the same holds for C. Next day the salesman will wake up either in D or E whence he can proceed to F or alternatively G at the end of the day. Then a single arrow links each of the latter with the final destination at H.

In this simple example it is only too easy to explore the whole collection of possible routes one by one; see Table 1.1. The shortest route from A to H takes thus 20 distance units (e.g. hundred miles); it involves visiting B, E, and F along the way.

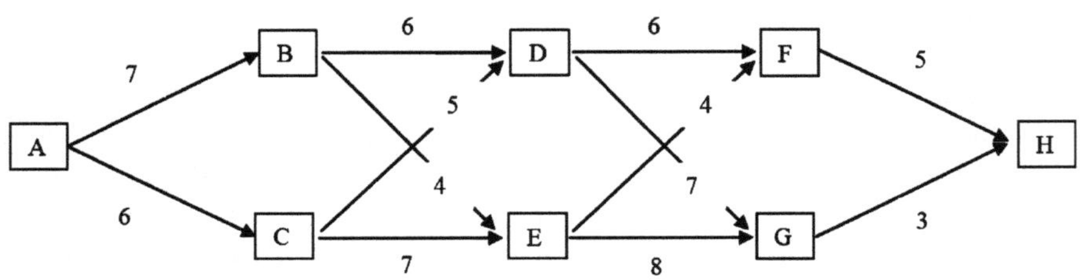

Fig. 1.1 A problem of choosing among routes

Table 1.1 Possible routes from city A to H

Possible routes	Total distance
A–B–D–F–H	$7 + 6 + 6 + 5 = 24$
A–B–D–G–H	$7 + 6 + 7 + 3 = 23$
A–B–E–F–H	$7 + 4 + 4 + 5 = 20$ min
A–B–E–G–H	$7 + 4 + 8 + 3 = 22$
A–C–E–G–H	$6 + 7 + 8 + 3 = 24$
A–C–E–F–H	$6 + 7 + 4 + 5 = 22$
A–C–D–F–H	$6 + 5 + 6 + 5 = 22$
A–C–D–G–H	$6 + 5 + 7 + 3 = 21$

Now, the proposed algorithm starts from destination and proceeds backward to the city of departure. Thus, from the point of view of the last daily trip, the most convenient place to depart from is city G; see Table 1.2. Nonetheless, whether G is along the optimal route or not remains to be seen. After all, this is not a one-day problem, so complete routes must be charted.

Let us go for the previous stage; Table 1.3. The salesman can start either at D or E. From D we can move on to F or G. The first route takes 11 to H while the second one takes only 10. Therefore, the best way from D to H goes definitely through G (and not F). The analysis runs similarly in case we depart from E on that day.

Let us go one stage backward. See Table 1.4. If we turn out to wake up at C we must choose between heading to D or E in our way to the final destination H. Comparing the two distances involved it is clear that the shortest path from C to H entails stopping at D (15 miles instead of 16). But note that right now we do not know if we shall depart from B or C on that day.

Going another stage backward we find ourselves at the very start of the trip. Again we have two options: from A we can go either to B or C, as shown in Table 1.5. The algorithm provides us the whole "payoff" of our "management strategy"; in our case, the distance of the routes that we have charted. The shortest path from A to H takes 20 miles; no other alternative is better. But there is more to this than meets the eye: the algorithm also charts the optimal itinerary to follow at any point in time. Thus, proceeding forward, from Table 1.5 by now we know that

Table 1.2 Relevant information for stage 4

City of departure	Decision: go to …	Distance to destination
F	H	5
G	H	3

Table 1.3 Relevant information for stage 3

City of departure	Decision: go to …	Distance to destination
D	F	$6 + 5 = 11$
	G	$7 + 3 = 10$ min
E	F	$4 + 5 = 9$ min
	G	$8 + 3 = 11$

Table 1.4 Relevant information for stage 2

City of departure	Decision: go to …	Distance to destination
B	D	6 +10 = 16
	E	4 + 9 = 13 min
C	D	5 + 10 = 15 min
	E	7 + 9 = 16

Table 1.5 Relevant information for stage 1

City of departure	Decision: go to …	Distance to destination
A	B	7 + 13 = 20 min
	C	6 + 15 = 21

we must go from A to B. Once we reach B, it is clear from Table 1.4 that we must head to E. Then, upon arrival to E, Table 1.3 suggests going to F. Last, Table 1.2 reminds us the distance from F to destination H. The cumulative distance along the route A–B–E–F–H is 7 + 4 + 4 + 5 = 20 (as we knew from Table 1.1).

A final, cautionary note is in order. It stresses the difference between step-by-step optimization and overall maximization. If we minimize the distance of each journey on a daily basis, following the information in Table 1.1 we would drive from A to C, since 6 miles is less than the 7 miles stretching from A to B. However, we have just seen that city C is not in the optimal route (which we have worked out by considering the overall problem). Therefore, it is important to keep a global vision of the whole issue; otherwise, myopic behavior can take us to apparent "shortcuts" that effectively lead us astray.

1.4.2 Maximizing Profit from Mineral Extraction

To illustrate the principle in a context closer to our interest we consider an ore mine with a certain remaining stock $Q(0)$, say 80 tons, right now ($t = 0$); this example draws heavily on Cerdá (2001). Management has a leasing contract which enables to continue exploitation for three more periods; after them the mine must be surrendered to the owners. Owing to the particular circumstances of the facility, there is a technical limit to the quantity extracted each period: the extraction level q can be either 0, 10 or 20 t. There is obviously a physical limit represented by the stock remaining at the time of the extraction. And there is also a rule governing the dynamics of the system: the stock at the end of any period is merely the difference between the stock at the beginning of that period and the extraction level in that period. At any time the profit to the firm depends inversely on the stock in the mine and directly on the quantity extracted; see Table 1.6. The problem is to determine the amount to extract each period so as to maximize cumulative profit over the whole period.

Table 1.6 Profit as a function of ore stock (Q) and extraction level (q)

	Q = 20	Q = 30	Q = 40	Q = 50	Q = 60	Q = 70	Q = 80
q = 0	0	0	−5	−5	−5	−10	−10
q = 10	−20	−15	5	10	20	25	30
q = 20	−30	−20	−10	−5	15	35	50

We first draw a diagram of the whole picture; see Fig. 1.2. Let t stand for the time at which the values of each variable are determined: $t = 0, 1, 2$. As already mentioned, $Q(t)$ denotes the remaining stock at the beginning of period t, while $q(t)$ is the extraction rate over period t.

Then we address the problem one period at a time. Crucially, we start from the last date and proceed backwards up to the initial time. We are particularly interested in the best course of action at each stage along with the cumulative profit from then on (Π). Since every decision thus takes account of the ensuing consequences we can be confident that all our decisions will indeed be optimal. This backward process will eventually take us to the very beginning where the same algorithm will guide us to the best decision. By then all the possible situations will have been assessed so we will get into chartered waters. In fact, we will have a complete map of the optimal route from the start to the finish.

End period: $\boldsymbol{Q}(3)$ **given** At the end of period 3 an amount $Q(3)$ of ore will be left in the mine. At that time, however, it is of no avail to us since we must give the mine back to its owners. Therefore, in terms of our objective function, there is no addition to the cumulative profit: the maximum "Profit till the end" is nil because we are already there: $\boldsymbol{\Pi}(3) = 0$.

This need not be the case in reality. Depending on the terms of the leasing contract managers can get some final reward or salvage value (e.g. a payment for the capital expenditures incurred, or the machinery, etc.). Or quite the opposite: perhaps management is obliged to incur some expenses to clear the site and restore it to its original appearance. In any case, should there be any payoff (whatever its sign) at this stage then it must be duly considered in the next round for overall maximization.

Period 2: $\boldsymbol{Q}(2)$ **given** Now the situation is quite different: we can manage the mine at will. Rational management involves trying to make the most of this (last) stage

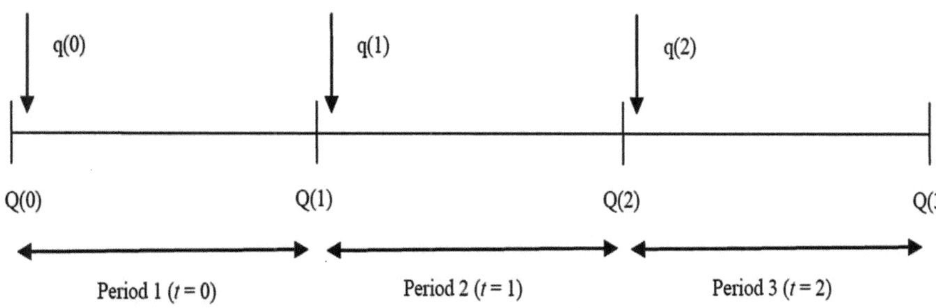

Fig. 1.2 State variables and decision variables in a three-period problem

$t = 2$. We know that the initial stock $Q(0)$ is 80 units. We also know that each period it is possible to extract either 0, 10, or 20 units. Consequently, by the time $t = 2$ the cumulative extraction level may have been: 0, 10, 20, 30, or 40 units. In other words, $Q(2)$ can take on the values: 80, 70, 60, 50, or 40. Whatever the actual level happens to be, the possible values of q remain the same. See Table 1.7.

For example, let $\boldsymbol{Q}(2) = 80$. The extraction level can alternatively be 0, 10 or 20. According to Table 1.6 the profit in each case amounts to -10, 30, and 50, respectively. Of course this current profit is important, but it is not the only one to take into account. Remember the above example: the distance between cities is crucial; but the optimization algorithm required us to compute the "Distance to destination". Similarly, here we are interested in the "Profit till the end". Therefore, these amounts must be augmented by any revenue (or cost) from the subsequent period. In our case, we have assumed it to be $\boldsymbol{\Pi}(3) = 0$, so the last column shows no addition; but again, this need not necessarily be the case. Further, this will no longer apply when we move back toward $t = 0$.

Next we synthesize the key results in a compact way as shown in Table 1.8. In particular, for $k = 2$ we list the optimal extraction level for each possible mineral stock and the maximum profit associated to them.**Period 1**: $\boldsymbol{Q}(1)$ **given** Once we have mapped the last period we proceed backward to the earlier one, $t = 1$. At this stage the mine has been exploited for only one period, so the stock can only take one of three possible values: 80, 70 or 60, depending on whether 0, 10 or 20 units have been extracted. See Table 1.9. Again we synthesize the relevant information in Table 1.10.

Once we have mapped this period we proceed backward to the earlier one which happens to be the initial one. The corresponding figures appear in Table 1.11.

Table 1.7 Relevant information for stage 2

Starting stock: $Q(2)$	Decision: extract $q(2)$	Profit till the end: $\Pi(2)$
80	0	−10
	10	30
	20	50 max
70	0	−10
	10	25
	20	35 max
60	0	−5
	10	20 max
	20	15
50	0	−5
	10	10 max
	20	−5
40	0	−5
	10	5 max
	20	−10

Table 1.8 Optimal roadmap for stage 2

Starting stock: $Q(2)$	Decision: extract $q^*(2)$	Profit till the end: $\Pi^*(2)$
80	20	50
70	20	35
60	10	20
50	10	10
40	10	5

Therefore, the maximum profit to be gained from optimally exploiting the mine in the two periods amounts to $\Pi^*(0) = 80$. There is no way to improve this mark. Furthermore, the optimal extraction path is neatly delineated. To this end, now we simply proceed forward. Following Table 1.6, first we take 20 units out:

$$q^*(0) = 20.$$

Since the initial stock is 80 units this means that the level at the end of the period (i.e. the starting level the next period) is:

$$Q^*(1) = Q(0) - q^*(0) = 80 - 20 = 60.$$

Going back to Table 1.5 we learn that for a stock this size the optimal extraction rate is $q^*(1) = 10$. Therefore the stock left for the next period is:

$$Q^*(2) = Q^*(1) - q^*(1) = 60 - 10 = 50.$$

And, according to Table 1.3, for this level of stock the best decision is to extract $q^*(2) = 10$. In the end, the stock left after the operation horizon turns out to be:

$$Q^*(3) = Q^*(2) - q^*(2) = 50 - 10 = 40.$$

We can easily check in Table 1.1 that following all these optimal decisions the maximum cumulative profit $\Pi^*(0)$ is certainly 80; it simply results from adding $50 + 20 + 10$ as shown in columns $Q = 80$, $Q = 60$, and $Q = 50$.

The above example is a (deterministic) optimization problem which can be formally stated as follows. A firm can produce a (say, monthly) amount of q units

Table 1.9 Relevant information for stage 1

Starting stock: $Q(1)$	Decision: extract $q(1)$	Profit till the end: $\Pi(1)$
80	0	$-10 + 50 = 40$
	10	$30 + 35 = 65$
	20	$50 + 20 = 70$ max
70	0	$-10 + 35 = 25$
	10	$25 + 20 = 45$ max
	20	$35 + 10 = 45$ max
60	0	$-5 + 20 = 15$
	10	$20 + 10 = 30$ max
	20	$15 + 5 = 20$

Table 1.10 Optimal roadmap for stage 1

Starting stock: $Q(1)$	Decision: extract $q^*(1)$	Profit till the end: $\Pi^*(1)$
80	20	70
70	10 or 20	45
60	10	30

Table 1.11 Relevant information for stage 0

Starting stock: $Q(0)$	Decision: extract $q(0)$	Profit till the end: $\Pi(0)$
80	0	$-10 + 70 = 60$
	10	$30 + 45 = 75$
	20	$50 + 30 = 80$ max

of output; this amount is typically bounded from above and below: $q_{min} \leq q \leq q_{max}$. Production at time t involves total costs which depend on the level of output q_t and the price of the input c_t; these costs are assumed to be:

$$C(q_t, c_t) = a_0 + a_1 c_t q_t + a_2 (q_t)^2.$$

Thus, there are fixed costs a_0 independent of the output level. There are also variable costs $a_1 c_t q_t$, where a_1 stands for the number of input units that are necessary to produce a unit of output; the term $a_2 (q_t)^2$ implies a rise in unit cost as q_t increases. Regarding total revenues, denoting the market price of the output by p_t, they are:

$$R(q_t, p_t) = p_t q_t.$$

At the beginning of each month the firm takes the prices c_t and p_t as given and solves an optimization program to choose the output level (the one that maximizes profits) subject to production constraints:

$$\text{Max}_{q_t} \Pi = p_t q_t - a_0 - a_1 c_t q_t - a_2 (q_t)^2$$

$$s.t.: q_{min} \leq q \leq q_{max}.$$

The first-order condition allows to derive the optimal output to produce:

$$p_t - a_1 c_t - 2a_2 q_t^* = 0 \rightarrow q_t^* = \frac{p_t - a_1 c_t}{2a_2}, \quad with\, q_{min} \leq q_t^* \leq q_{max}.$$

Needless to say, in practice both c_t and p_t evolve over time in an unpredictable way. Choosing the optimal output level then becomes a stochastic optimal control problem. We will have more to say on these problems in the coming chapters; in particular, we will propose some theoretical models for the behavior of prices, and develop some numerical methods to come up with quantitative solutions to these problems.

1.4.3 A Rigorous Exposition

The above problem and solution algorithm can be stated in a quite formal, rigorous way. Next we describe the main constituents to the problem while indentifying them in the above examples. Note that we are interested in optimally managing a project over time. Thus we deal with a dynamic system. The system as such can be described at any time by a state variable. This variable changes over time according to a given equation; one key ingredient here is the decision variable that we fully control (subject to certain constraints), which bears a direct impact on the system dynamics. Last, but certainly not least, we manage the control variable so as to optimize an objective function over a given time frame.

Our two examples are stated in discrete time: it makes full sense to talk about time t and time $t + 1$. The number of periods or stages N is known from the very outset. In the route example the salesman (or the tank wagon driver) will spend 4 days on the road. In the case of the mine, management commands full control over extraction for three periods.

At $t = 0$, the initial state or starting point of the system $x(0)$ is given exogenously. We just find ourselves at city A, or are endowed with 80 tons of mineral ore. The state evolves over time according to the value that we assign to our control variables $u(t)$: it can be the length of a daily trip, or the amount of yearly extraction. Since we have $t \in \{0, 1, 2, \ldots, N - 1\}$, we naturally have $u = \{u(0), u(1), u(2), \ldots, u(N - 1)\}$. Note that the control variables are confined to taking on a limited set of possible values $\Omega(t)$: the salesman's car cannot possibly go from A to H in a single day, just as the miners operate with available machinery inside tunnels of a given size (so they can only extract 0, 10, or 20 tons). Obviously, there is also a set of state variables $x(t)$ ranging from the initial value $x(0)$ to the final value $x(N)$.

The state variables and the control variables are related through a system of difference equations f (the state equations); for each t, they basically describe the end state as a function of the starting state, the control adopted, and t itself. In the first example, on the second day our salesman will arrive at city D if he/she departed from C and drove over 5 miles. Similarly, the mine will contain 60 tons at the end of period 2 if there were 70 at the beginning of that period and 10 tons have been removed in the meantime.

Now, the decision maker controls the system so as to optimize the objective function F. For example, this can be to minimize expenses or maximize profits. Starting at $t = 0$ from $x(0)$ it is necessary to choose $u(0) \in \Omega(0)$. At this stage there is a contribution to the objective function worth $F[x(0), u(0), 0]$. Hence we move on to $t = 1$ when the state variable takes on the value $x(1) = f(x(0), u(0), 0)$. Again, a control must be chosen from an admissible set, and a new contribution to the objective function takes place. This process is repeated until $t = N - 1$. The system starts from $x(k - 1)$; managers choose $u(t - 1) \in \Omega\,(t - 1)$, thus contributing $F[x(N - 1), u(N - 1), N - 1]$ to the objective function. The system eventually reaches the final state $x(N)$ which entails a terminal contribution $S[x(N)]$ (it can be positive in the case of a salvage value, or negative when there are costs to decommissioning the facilities). Note that all the cash flows in this program are equally weighted irrespective of their specific dates (i.e. there is no discounting or, equivalently, the discount rate is assumed to be zero).

In sum, the problem can be analytically stated as:

$$\begin{aligned} \max_{\{u(t)\}_{t=0}^{t=N-1}} J &= \sum_{t=0}^{t=N-1} F[x(t), u(t), t] + S[x(N)], \\ s.t.: x(t+1) &= f(x(t), u(t), t), \\ x(0) &= x_0, \\ u(t) &\in \Omega(t). \end{aligned} \tag{1.23}$$

Dynamic programming provides a solution. Let $J^*(x_0)$ be the optimal value of the objective function (20 distance units and 80 t in the above examples, respectively). Then,

$$J^*(x_0) = J_0^*\{x_0\}.$$

The function $J_0^*\{x_0\}$ is given by the last stage of the following algorithm, which starts at the end of the time horizon and proceeds backward to the beginning:

$$J_N^*\{x_N\} = S[x(N)]; \tag{1.24}$$

and for $t \in \{N-1, N-2, \ldots, 1, 0\}$:

$$J_t^*\{x_t\} = \max_{u(t)\in\Omega(t)} \{F[x(t), u(t), t] + J_{t+1}^*\{x(t+1)\}\}. \tag{1.25}$$

The set Eqs. (1.22)–(1.25) comprises the so-called Bellman equations for the problem at hand. Besides, if $u^*(t)$ maximizes the expression in the right hand of the Bellman equation, as a function of $x(t)$ for each $t \in \{0, 1, 2, \ldots, N-1\}$, the vector

$$u^* = (u^*(0), u^*(1), u^*(2), \ldots, u^*(N-1)) \tag{1.26}$$

is the optimal control of the problem. In the salesman example, the optimal route turned out to be A–B–E–F–H. In the case of the mine, the sequence of optimal extraction rates was (20, 10, 10).

1.5 Where Next?

We have just seen that putting a price on time (or dynamic cash flows) is far from obvious. The next chapter opens the door to uncertainty. This will obviously complicate matters quite a lot. Nonetheless, markets have been putting a price on uncertainty for a long while. As an example, assume that there is zero-coupon bond issued by the German Treasury which promises to pay €100 in 3 years' time. At the same time, there is also a bond issued by the Spanish Treasury which promises the same amount at the same time. Despite their similarities, the first one today changes hands at €97 while the second trades for €89. In principle we could think in terms of discounted cash flows:

$$97 = \frac{100}{(1 + \mathrm{r_G})^3}; \quad 89 = \frac{100}{(1 + \mathrm{r_S})^3}.$$

The different valuations imply different discount rates r_G and r_S for these German and Spanish bonds, respectively, since all the other parameters of the bonds involved are equal.

The rates r_G and r_S for Germany and Spain can be conceptually broken down into two components. The first one must reward investors for the simple fact that they must wait (i.e. defer their consumption) up to 3 years (even if the bonds were actually risk free). There is no obvious reason to think that investors in the two bonds differ in their patience, so this component can well be the same (r). The second component, instead, must reward them for assuming risk. Since the two bonds differ in perceived risk (€97 vs. €89), investors are applying a different risk premium to each bond (denoted by g and s, respectively). We thus have $r_G \equiv \mathrm{r} + \mathrm{g}$ and $r_S \equiv \mathrm{r} + \mathrm{s}$. If the assumed prices (€97 and €89) were actual market prices then we could get an estimate of the risk premia g and s as perceived by the markets (for that particular maturity, 3 years).

Note that the risk premium can differ for other expiration dates. Thus, actual data in August 2012 showed that the 3-year government bond yield was 3.88 % and 0.13 % for Spain and Germany, respectively; hence the (3-year) differential return was 3.75 %. However, 10-year bonds yielded 6.58 and 1.34 %, respectively, so the (10-year) differential return amounted to 5.24 % (this is the 'risk premium' usually quoted in the media). Fortunately, in addition to markets there is also a sizeable core of proven concepts and methodologies to grapple with uncertainty effectively in many instances.

References

Bellman R (1957) Dynamic programming. Princeton University Press, Princeton

Cerdá E (2001) Optimización dinámica. Pearson Educación, Madrid

de Neufville R, Scholtes S (2011) Flexibility in engineering design. The MIT Press, Cambridge, MA

Fisher I (1907) The rate of interest: its nature, determination, and relation to economic phenomena. Macmillan, New York

Fisher I (1930) The theory of interest. Macmillan, New York

Luenberger DG (2009) Investment science. Oxford University Press, New York

Part II
Investment Under Uncertainty

… in this world nothing can be said to be certain, except death and taxes. Benjamin Franklin (1706–1790).

Part II
Investment Under Uncertainty

Chapter 2
Theoretical Foundations

2.1 Mean–Variance Analysis in a Single Period

In this section we address the portfolio selection problem faced by an individual investor. In the absence of uncertainty (as in Chap. 1), the problem of choosing among assets can be fully characterized by means of a single measure of investment performance, namely return (with the distinction between expected return and realized return ringing hollow). Under uncertainty, however, it is simply impossible to perfectly forecast the outcome of a given decision. The potential results are typically described by a frequency distribution. If the possible returns on a given asset are just a few then it is sensible to set up a table with the whole list of possibilities in a column and their corresponding probabilities on the other. If, instead, the number of possibilities is very high the above approach becomes too complex. Capturing the relevant information in the frequency distribution requires more than one descriptive statistic. The standard approach rests on using just two: a measure of position (the average), and a measure of dispersion around it (the variance or the standard deviation). These two measures can suffice to describe all the randomness in the distribution of uncertain returns in some special cases, e.g. when they are normally distributed. However, these statistics are clearly insufficient in other cases, for instance if the distribution of returns is asymmetric or displays fat tails. Under these circumstances, additional statistics (to gauge skewness, kurtosis, etc.) are required.

2.1.1 Characteristics of Asset Returns

The average outcome The mean or average value of a distribution is sometimes referred to as the mathematical expectation. Consider a set of risky assets denoted by i, with $i = 1, 2, \ldots, n$. The *ex post* return R of any asset is unknown *ex ante*, since it will depend on the state of nature prevailing at that time and this is impossible to forecast exactly. The sub-index j, with $j = 1, 2, \ldots, m$, stands for

L.M. Abadie and J.M. Chamorro, *Investment in Energy Assets Under Uncertainty*, Lecture Notes in Energy 21, DOI: 10.1007/978-1-4471-5592-8_2,

each state of nature. Thus, the number of possible returns for any asset equals m. The return on asset i in state j is therefore represented by R_{ij}; its probability of occurrence is P_{ij}. The expected return is alternatively denoted by $E(R_{ij})$ or $\overline{\boldsymbol{R}_i}$:

$$\overline{\boldsymbol{R}_i} \equiv E(R_{ij}) = \sum_{j=1}^{m} P_{ij} R_{ij}. \tag{2.1}$$

A useful property of the mathematical expectation is that the expected value of the sum of two returns equals the sum of their expected returns:

$$E(R_{1j} + R_{2j}) = \overline{R_1} + \overline{R_2}. \tag{2.2}$$

Another property is that the expected value of a constant c times an uncertain return equals the constant times the expected return:

$$E(cR_{1j}) = c\overline{R_1}. \tag{2.3}$$

A measure of dispersion Whenever a random variable can take on a number of possible results it is convenient to learn how much they deviate from the average. The usual measure is the variance. All the deviations are squared (to avoid any offsetting) and then weighted by their respective probabilities:

$$\sigma_i^2 = \sum_{j=1}^{m} P_{ij} \left(R_{ij} - \overline{R_i}\right)^2. \tag{2.4}$$

The square root of the variance is the standard deviation or volatility.

Box 2.1 Example with two risky assets
Below we show the end-of-month prices of stocks A and B. We want to compute their average or expected returns and standard deviations.

Month	A (price)	B (price)	A (return)	B (return)
0	25	45		
1	24.125	44.875	−0.035	−0,002
2	23.375	46.875	−0.031	0,044
3	24.75	45.25	0.058	−0,034
4	26.625	50.875	0.075	0,124
5	26.5	58.5	−0.004	0,149
6	28	57.25	0.056	−0,021
7	28.875	62.75	0.031	0,096
8	29.75	65.5	0.030	0,043
9	31.375	74.375	0.054	0,135
10	36.25	78.5	0.155	0,055
11	37.125	78	0.024	−0,006
12	36.875	78.125	−0.006	0,001

The starting point is computing (monthly) returns. The standard formula for the rate of return is simply $R(t) = [\mathrm{P(t)} - \mathrm{P(t-1)}]/\mathrm{P(t-1)}$. Thus, the first monthly return on asset A is $(24.125 - 25)/25 = -0.0350$, i.e. a drop of 3.5 %. Alternatively, the logarithmic return is frequently used. It is computed as the natural logarithm of the price relative: $\ln[\mathrm{P(t)}/\mathrm{P(t-1)}]$; in the above case we get: $\ln\left[\frac{24.125}{25}\right] = \ln 0.965 = -0.0356$. Similarly for asset B. Upon computation of asset returns we compute their expectation by simply averaging them. We get: $\overline{R_A} = 0.034$ and $\overline{R_B} = 0.049$, respectively.

As for the standard deviations, we first subtract $\overline{R_A}$ from each return on A, square the resulting differences, add them all up, and divide by 12. Analogously for B. We get the variances: $\sigma_A^2 = 0.0025$ and $\sigma_B^2 = 0.0038$, respectively. There is an alternative way to compute the variance of returns: first, all the returns are squared and then aggregated; next we subtract the average return squared (any handbook on descriptive statistics proves this correct). The standard deviations are just the square root of these values. Usually annualized volatilities are used, so we multiply these (monthly) figures by $\sqrt{12}$; this yields: $\sigma_A = 0.174$ and $\sigma_B = 0.213$, .or 17.4 % and 21.3 % , respectively.

If the returns on asset A and B actually have a normal distribution, the above returns are twelve independent samples from a normal distribution, and their standard deviation is the monthly volatility of each asset. Obviously, different samples from the same distributions would give different estimates of σ_A and σ_B. To what extent can we be sure that our estimator has desirable properties? In fact, it turns out to be biased; fortunately, however, this is easy to correct. The correction boils down to divide the sum of square deviations by the number of observations less one (in our case, 11). Thus, the unbiased variance estimates are $\sigma_A^2 = 0.0028$ and $\sigma_B^2 = 0.0041$; of course, the volatility estimates change accordingly: $\sigma_A = 0.182$ and $\sigma_B = 0.223$. In sum, unbiased annualized volatility is computed as:

$$\sigma_i^2 = \frac{1}{M-1}\frac{1}{\Delta t}\sum_{j=1}^{M}\left(R_{ij} - \overline{R_i}\right)^2,$$

where M is the number of returns available, and Δt refers to the time interval in annual terms (in the example, 1/12).

2.1.2 Characteristics of Portfolio Returns

Consider a portfolio P consisting of N individual assets ($i = 1, 2, \ldots, N$, with $N \leq n$) in certain proportions X_i:

$$P = X_1R_1 + X_2R_2 + \cdots + X_NR_N, \; with \sum_i X_i = 1. \tag{2.5}$$

The return on P in state j is the weighted average of the asset returns in that state:

$$R_{Pj} = \sum_{i=1}^{N} X_i R_{ij}. \tag{2.6}$$

The above two properties of the mathematical expectation allow compute the expected return as a weighted average of expected returns:

$$\overline{R_P} \equiv E(R_{Pj}) = \sum_{i=1}^{N} X_i \bar{R}_i. \tag{2.7}$$

The variance of P, however, is not merely a sum of variances. To see this, consider the case of a portfolio comprising just two assets: $i = 1, 2$. We know from Statistics that:

$$\sigma^2(\tilde{X} + \tilde{Y}) = \sigma_X^2 + \sigma_Y^2 + \sigma_{XY} + \sigma_{YX}.$$

Here σ_{XY} denotes the covariance between the random variables X and Y. It provides a measure of the extent to which they tend to move together. Of course, $\sigma_{XY} = \sigma_{YX}$. In our case the variance of a two-asset portfolio is:

$$\begin{aligned} \sigma_P^2 &= E\left(R_{Pj} - \overline{R_P}\right)^2 = X_1^2\sigma_1^2 + X_2^2\sigma_2^2 + X_1X_2\sigma_{12} + X_2X_1\sigma_{21} \\ &= X_1^2\sigma_1^2 + X_2^2\sigma_2^2 + 2X_1X_2\sigma_{12} = X_1^2\sigma_1^2 + X_2^2\sigma_2^2 + 2X_1X_2\rho_{12}\sigma_1\sigma_2. \end{aligned} \tag{2.8}$$

Here σ_{12} denotes the covariance between the uncertain returns on asset 1 and asset 2. It will be large and positive if assets 1 and 2 tend to perform above average (or below it) at similar times. If positive and negative deviations (with respect to their averages) are unrelated for the most part then σ_{12} will approach zero. And if the two assets tend to move in opposite directions the covariance will be negative. A related measure of co-movement is the correlation coefficient ρ_{12}:

$$\rho_{12} = \frac{\sigma_{12}}{\sigma_1\sigma_2}, \tag{2.9}$$

the value of which ranges between -1 (perfect negative correlation) and +1 (perfect positive correlation).

If we know the expected returns of assets 1 and 2 alongside their standard deviations and correlation coefficient then we can trace the locus over which all their possible combinations lay in the $\bar{R} - \sigma$ space.

Box 2.2 Portfolio frontier with two risky assets

Following with the assets in Box 2.1 above, now we want to look at the characteristics of their combinations in more detail. A key element here is the covariance between their returns (or the correlation coefficient).

$R_A(t)$	$R_B(t)$	$R_A(t) - \overline{R_A}$	$R_B(t) - \overline{R_B}$	Product
−0.035	−0.003	−0.069	−0.052	0.004
−0.031	0.045	−0.065	−0.004	0.000
0.059	−0.035	0.025	−0.084	−0.002
0.076	0.124	0.042	0.075	0.003
−0.005	0.150	−0.039	0.101	−0.004
0.057	−0.021	0.022	−0.070	−0.002
0.031	0.096	−0.003	0.047	0.000
0.030	0.044	−0.004	−0.005	0.000
0.055	0.135	0.021	0.087	0.002
0.155	0.055	0.121	0.007	0.001
0.024	−0.006	−0.010	−0.055	0.001
−0.007	0.002	−0.041	−0.047	0.002

Now averaging all the cross products we get the covariance between A and B: $\sigma_{AB} = 0.00036$. The correlation coefficient provides a normalized measure of the extent to which they change with respect to each other; in our case: $\rho_{AB} = 0.00036/[(0.050)(0.061)] = 0.1175$.

Now we want to plot the combinations of these two risky assets in the expected return—standard deviation space. These portfolios consist only of A and B, with both $0 \leq X_A \leq 1$ and $0 \leq X_B \leq 1$; besides, these weights sum to one: $X_A + X_B = 1$, so $X_B = (1 - X_A)$. For each weight, say $X_A = 0.05$, the expected return is given by Eq. (2.7); the standard deviation is computed as the square root of Eq. (2.8). The extremes of the frontier are represented by assets A and B, respectively. See Fig. 2.1.

If it is possible to sell one of the assets short and invest the revenue in the other asset, then $X_A < 0$ and $X_B = (1 - X_A) > 1$. In this case, the portfolio frontier (in fact, a hyperbola) extends way beyond A and B.

Equation (2.8) is a quadratic expression, so in general the combinations fall over a curve. The left-most combination is called the minimum variance portfolio, *mvp*. It can be proven that the part of the curve above the *mvp* and to the right is

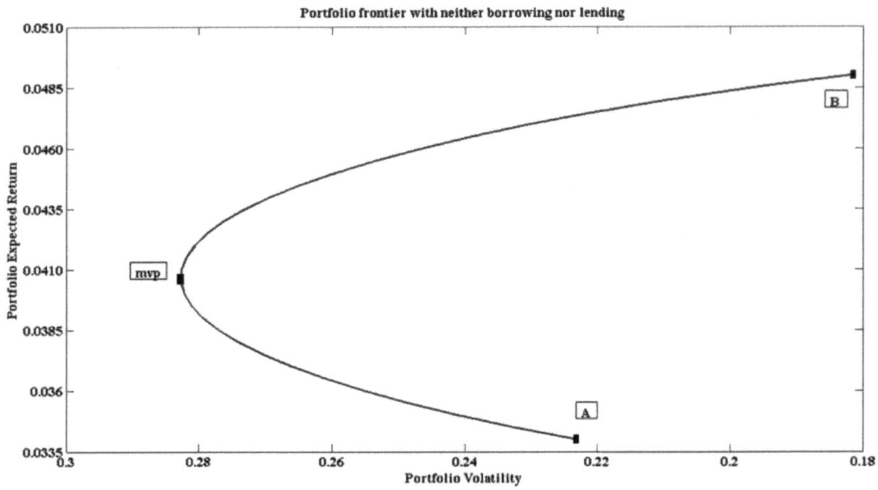

Fig. 2.1 Portfolio frontier with neither borrowing nor lending ($0 \leq X_i \leq 1, \sum_i X_i = 1$)

convex, while the part below it is concave. Investors naturally like (expected) return. Therefore, given two portfolios with the same σ they opt for the portfolio in the upper part. Since every combination in the lower part has a mirror image in the upper part, all the combinations in the former are dominated by others in the latter. At the same time, (risk averse) investors dislike risk. Therefore, given two portfolios with the same $\overline{R}$ they opt for the portfolio that is most to the left. This two forces imply that the set of dominating portfolios (the so-called efficient frontier) starts at the *mvp* and goes up and to the right in a concave pattern.

When the portfolio includes a number N of assets the variance is thus:

$$\sigma_P^2 = \sum_{i=1}^{N} X_i^2 \sigma_i^2 + \sum_{i=1}^{N} \sum_{\substack{j=1 \\ j \neq 1}}^{N} X_i X_j \sigma_{ij}. \tag{2.10}$$

Assume that all the assets in the portfolio have equal weights: $X_i = X_j = 1/N$. Then Eq. (2.10) becomes:

$$\sigma_P^2 = \sum_{i=1}^{N} \left(\frac{1}{N}\right)^2 \sigma_i^2 + \sum_{i=1}^{N} \sum_{\substack{j=1 \\ j \neq 1}}^{N} \left(\frac{1}{N}\right)\left(\frac{1}{N}\right) \sigma_{ij}. \tag{2.11}$$

This expression can be equivalently stated as:

$$\sigma_P^2 = \frac{1}{N} \sum_{i=1}^{N} \frac{1}{N} \sigma_i^2 + \frac{N-1}{N} \sum_{i=1}^{N} \sum_{\substack{j=1 \\ j \neq 1}}^{N} \frac{1}{N(N-1)} \sigma_{ij}. \tag{2.12}$$

The first term in the right hand is 1/N times the sum of all the variances σ_i^2 divided by their number N (i.e. the average variance). Similarly, the second term is $(N-1)/N$ times the sum of all the covariances σ_{ij} divided by their number $N(N-1)$ (i.e. the average covariance). In sum, the variance of the portfolio is:

$$\sigma_P^2 = \frac{1}{N}\overline{\sigma_i^2} + \frac{N-1}{N}\overline{\sigma_{ij}}. \tag{2.13}$$

As the first term suggests, individual risks (as measured by the variance) can be diversified away by adding new assets into the portfolio; indeed, this term approaches zero as N tends to infinity. The second term, however, remains: the average covariance between asset returns constitutes the lowest possible bound. The contribution of covariance terms thus cannot be eliminated through diversification. See Markowitz (1987a, b).

2.1.3 Riskless Borrowing and Lending

Up to now we have considered all the available assets $i = 1, 2, \ldots, n$ as risky ($\sigma > 0$). The inclusion of a risk-free asset or security F not only adds realism but simplifies analysis as well. Lending at the riskless rate can be interpreted as purchasing an asset F with a certain return r (assumed constant). And selling this asset amounts to borrowing at the risk-free rate r. Since F is a safe asset its return variance is zero ($\sigma_F = 0$) and also its covariance with any other asset ($\sigma_{iF} = 0$, for all i).

Assume that investors are allowed to borrow and lend unlimited amounts of money at the riskless rate r. An individual investor has an initial wealth to be invested in the risk-free asset F and a risky portfolio G in proportions $(1 - f)$ and f, respectively. Given how investors are assumed to behave, G is anticipated to be in the efficient frontier. The expected return of this portfolio is:

$$\overline{R} = (1 - f)r + f\overline{R_G}, \tag{2.14}$$

where $\overline{R_G}$ denotes the expected return on portfolio G. The variance of this portfolio is:

$$\sigma^2 = (1 - f)^2\sigma_F^2 + f^2\sigma_G^2 + 2f(1 - f)\sigma_{FG}, \tag{2.15}$$

where σ_G^2 stands for the variance of G, and σ_{FG} for its covariance with asset F. Given the safe character of F, this expression can be simplified to yield:

$$\sigma^2 = f^2\sigma_G^2 \rightarrow \sigma = f\sigma_G. \tag{2.16}$$

Solving for f here ($f = \sigma/\sigma_G$) and substituting in Eq. (2.14) we get:

$$\overline{R} = \left(1 - \frac{\sigma}{\sigma_G}\right)r + \frac{\sigma}{\sigma_G}\overline{R_G} = r + \frac{\overline{R_G} - r}{\sigma_G}\sigma \tag{2.17}$$

This is the equation of a straight line in the standard deviation—expected return space. The intercept on the vertical axis $\overline{R}$ is the riskless rate r. The slope is the ratio of the excess return of portfolio G over the safe asset F and the standard deviation of portfolio G; it represents the extra return on G per unit of risk (as gauged by its volatility). The sign of this slope depends on investors' behavior toward risk.

There are three possible attitudes. They are most intuitively defined by using a fair gamble. Assume there is a lottery with two possible payoffs, \$2 and zero, both equally probable (it could also be an oil rig tapping on a well which can turn out to be huge or pretty small). The expected payoff from the lottery is:

$$E(payoff) = \frac{1}{2}0 + \frac{1}{2}2 = \$1.$$

If this lottery is to be fair then its cost must equal the expected value, \$1. Therefore, if the investor forgoes the lottery then she keeps the \$1 note in the pocket. We thus have two alternatives or options: "invest" (and earn \$1 on average) and "do not invest" (earn \$1 for sure). Now, an investor is said to be risk neutral if she is indifferent between playing the lottery and foregoing it. If the investor rejects the fair gamble then she is said to show risk aversion. And if she strictly prefers to play this lottery then she is a risk lover or shows risk seeking.

We assume that overall investors are risk averse. This does not mean that they will never undertake risky projects. They will accept risk but only if they are duly rewarded for it. Turning back to the example of the lottery, a risk averse investor will reject the lottery as it is. However, she can well accept it if the best payoff rises from \$2 to \$4 (in which case the expected payoff is twice as high as the cost). Perhaps another investor will only accept the gamble if the top prize reaches \$6. In sum, risk aversion can be more or less intense, or there can be different degrees of risk aversion. From the viewpoint of the asset market, the average taste for risk of the universe of investors is the relevant issue. If risk aversion is the dominant profile and the (risky) portfolio G has a place for in the market then its expected return $\overline{R_G}$ must be higher than the riskless rate r. Therefore, the slope of the straight line $(\overline{R_G} - r)/\sigma_G$ is positive: $\overline{R}$ grows linearly with σ.

There is nothing peculiar to portfolio G; it is a generic (efficient) portfolio. Now look at Fig. 2.2. Confronted with all the possibilities along the efficient frontier, the typical investor will seek the risky portfolio which (combined with F) allows her to go as high as possible for a given level of risk; let H denote this much sought-after risky portfolio. The highest possible straight line starts from the intercept at r. Its slope will be $\frac{\overline{R_H} - r}{\sigma_H} > 0$.

Note that the identity or precise composition of portfolio H does not depend on investors' preferences. As long as they are risk averse they just like return and dislike risk. The only parameters required for identifying H are expected returns and standard deviations. If this information is shared by all investors, then all of them frame the portfolio selection problem in similar terms and come up with the same solution.

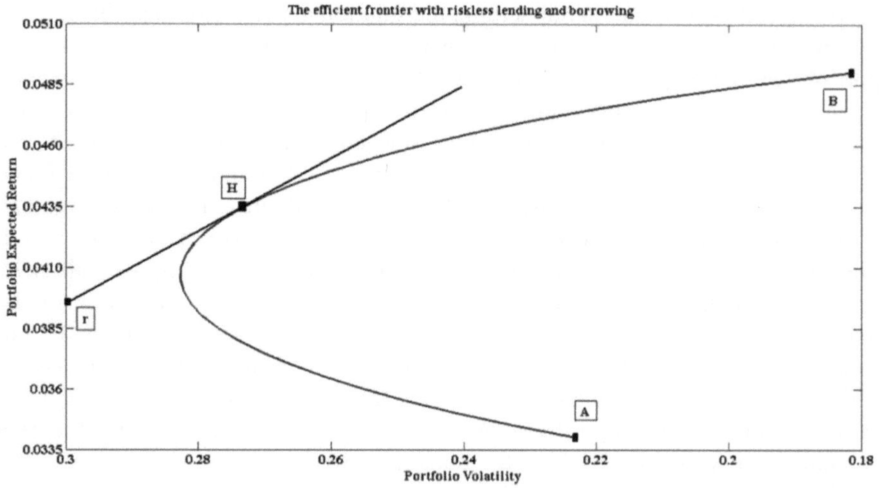

Fig. 2.2 The efficient frontier with riskless lending and borrowing

Then, what is the role played by investors' taste for risk? Each investor will combine *H* with the safe asset *F* according to her preferences. For example, an aggressive investor can place 99 % of her wealth in *H* and only 1 % in *F*. Instead, another investor can go for 99 % in *F* and only 1 % in *H*. Still another can borrow at rate *r* and invest her wealth alongside these revenues in *H*. The key point here is that *H* satisfies all investors alike when it comes to the best risky portfolio in presence of a riskless asset. Thus *H* constitutes the solution to their portfolio selection problem.

2.2 The Standard Capital Asset Pricing Model

The above section dealt with the choice of the optimal portfolio by an individual or institutional investor. This section instead introduces a pricing model. Thus we are interested in the determination of market prices for risky assets and portfolios of assets. Obviously markets are driven by the aggregate of investors. How does this community behave? As long as any single investor behaves as explained before we can extrapolate and build a general equilibrium model. The first and most simple such model is the standard Capital Asset Pricing Model that we sketch below.

Assume a frictionless, perfectly competitive capital market. All investors adopt the mean–variance approach. Thus they only care about expected returns and standard deviations. They are further assumed to share the same probabilistic beliefs regarding all the assets in the market, and to define the investment horizon in the same way. Last, they can borrow or lend unlimited amounts at the riskless interest rate.

We already know that, in this setting, the solution to the portfolio problem faced by any investor rests on two pillars, namely the risk-free asset F and a risky portfolio H that is the best candidate (among all risky portfolios) to combine with F. Now, if all the investors frame the portfolio problem in exactly the same terms then all of them will come up with the same ingredients to the solution: F and H. Of course, different appetites for risk will lead to different proportions of F and H in individual portfolios, but the basic fact remains: all are composed of F and H.

Since the risky portfolio H is the same across the universe of investors then it must mirror the market portfolio, M. The market portfolio comprises all risky assets. Given that all the individuals hold the same H, the composition of the market portfolio exactly matches the composition of H, i.e. M mimics H. For example, if $N = 2$ and the optimal H for each investor consists of 20 % in asset 1 and 80 % in asset 2 then the composition of M is [0.20; 0.80]. Of course, though, M dwarfs any individual portfolio. Further, asset 1 will take 20 % of total market capitalization with asset 2 taking the remaining 80 %.

Now, substituting M for H in Eq. (2.17) yields:

$$\overline{R} = r + \frac{\overline{R}_M - r}{\sigma_M}\sigma. \tag{2.18}$$

This is again the equation of a straight line but a special one, the capital market line. As before, it starts from the riskless rate on the $\overline{R}$ axis. The slope now is $\frac{\overline{R_M - r}}{\sigma_M}$, which is the market price of risk. Thus, if the excess return on the market is 0.06 and $\sigma_M = 0.2$ then $\frac{\overline{R_M - r}}{\sigma_M} = 0.3$. Therefore, if the volatility of a (efficient) portfolio increases by 1 % point, the expected return on the portfolio must increase by 30 basis points (as seen by the market).

Equation (2.18) is certainly useful as long as we assess an efficient portfolio. But we can face portfolios whose efficient character we are not aware of. And, of course, we can also ponder at some time the decision to purchase (or sell) particular stocks. Next we show an expression for the (expected) risk-return relationship in equilibrium for whatever asset or portfolio of assets. We omit the proof; the interested reader can look it up in a number of references such as Sharpe (1964), Elton et al. (2009) or Luenberger (2009).

As explained above, we take it for granted that the market portfolio M is an efficient portfolio. According to the Capital Asset Pricing Model (CAPM), the expected return of any asset i (or portfolio) is a linear function of its risk:

$$\overline{R_i} = r + \beta_i\left(\overline{R_M} - r\right), \tag{2.19}$$

where $\beta_i = \frac{\sigma_{iM}}{\sigma_M^2}$. This is the security market line; it applies to any security or portfolio.

Just as the difference $\overline{R_M} - r$ is the expected excess return on the market portfolio, $\overline{R_i} - r$ is the expected excess return on asset i. The latter turns out to be β_i times the former in equilibrium. This coefficient is called the beta of the asset. It is a measure of the asset's risk relative to the market: σ_{iM}/σ_M^2. Is this actually the relevant measure of the asset risk? Yes. All the investors put a fraction of their wealth in M; given that this is the market portfolio, M is a diversified portfolio, indeed the most diversified possible. This in turn implies that diversification has absolutely operated its magic in terms of reducing portfolio risk. Specifically, idiosyncratic risks have been made to vanish. Yet the extent to which security returns respond to market changes (the non-diversifiable or systematic risk) remains intact; this component of risk cannot be similarly pushed to extinction by enlarging the portfolio since this is already the market portfolio.

Box 2.3 Computing beta

Let there be a huge number N of traded assets. Assume that we know the return on two particular assets, 1 and 2, along with that on the market portfolio, as gauged by an index M, over the last 6 months.

Month	R_{1t}	R_{2t}	R_{Mt}
1	10	0	4
2	3	5	2
3	15	7	8
4	9	13	6
5	3	1	0
6	2	4	4

From these observed data we want to estimate β_1 and β_2. We start computing the average returns and the variances:

$$\overline{R_1} = \frac{42}{6} = 7; \sigma_1^2 = \frac{134}{6}; \overline{R_2} = \frac{30}{6} = 5; \sigma_2^2 = \frac{110}{6}; \overline{R_M} = \frac{24}{6} = 4; \sigma_M^2 = \frac{40}{6}.$$

Next we derive the covariances with the market index.

$$\sigma_{1M} = \frac{1}{6}\begin{bmatrix} (10-7)(4-4)+(3-7)(2-4)+(15-7)(8-4) \\ +(9-7)(6-4)+(3-7)(0-4)+(2-7)(4-4) \end{bmatrix} = \frac{60}{6};$$

$$\sigma_{2M} = \frac{1}{6}\begin{bmatrix} (0-5)(4-4)+(5-5)(2-4)+(7-5)(8-4) \\ +(13-5)(6-4)+(1-5)(0-4)+(4-5)(4-4) \end{bmatrix} = \frac{40}{6}.$$

Hence we get the betas.

$$\beta_1 = \frac{\sigma_{1M}}{\sigma_M^2} = \frac{60/6}{40/6} = 1.5;\ \beta_2 = \frac{\sigma_{2M}}{\sigma_M^2} = \frac{40/6}{40/6} = 1.$$

Figure 2.3 displays 60 monthly returns on a stock (Exxon) and a market index (S&P 500). It also shows the regression line that best fits the data; the slope is just beta.

It is easy to draw the security market line in the $\overline{R} - \beta$ space. Note that just two points are enough for this. Thinking in terms of the intercept, we can choose the safe asset F, the beta of which is zero (it bears no systematic risk). According to Eq. (2.19), when $\beta = 0$ the asset must earn r in equilibrium. Another point is given by the market portfolio itself: $\beta_M = \sigma_{MM}/\sigma_M^2 = 1$. In this case, the CAPM implies that it must earn $\overline{R_M}$. Once we have drawn this straight line, we can take the beta of any asset onto the horizontal axis and derive its expected return (according to the CAPM).

Should we have a different expectation then we could buy or sell the asset accordingly (provided we trust in the estimated CAPM). This is owing to the fact that an asset lying above or below the security market line involves an arbitrage opportunity, i.e. the chance to get a free, sure profit. To see it, assume that we have estimated the relationship $\overline{R_i} = 2.2 + \beta_i 2$. Let C denote a security above the line; assume its beta is 1.4. If, according to the estimated CAPM, it should earn 5 % and it is actually earning 5.5 % then we should sell a portfolio D short (with the same beta 1.4) but located on the CAPM line (thus, a combination of F and M). If we are right, a $1,000 from the sale of D and invested in C will earn us a $5 profit without bearing any risk (while our own wealth remains at bay). These opportunities can indeed arise, but they will be quite short-lived for their very nature. The notion of market equilibrium is inconsistent with any such arbitrage opportunity.

We have stated the CAPM in terms of expected returns. This is the standard practice. Some people may feel confused by the P of "pricing" in the model name, in that it suggests amounts expressed in monetary units. However, there is no room

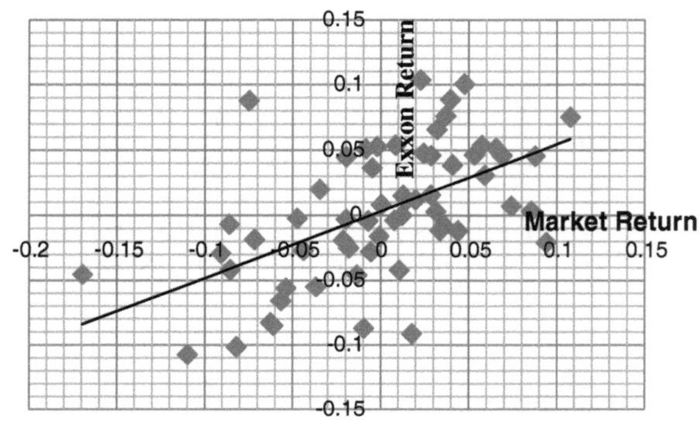

Fig. 2.3 Scatter diagram of returns on Exxon and SP&500 over 60 months

for confusion here. We can rewrite this equilibrium relationship in terms of the asset's current and expected prices using a bit of algebra. We do not pursue this issue further because we will not need it in the forthcoming chapters. Again, interested readers can resort to the references at the end of the chapter.

2.3 Single-Period Risk-Neutral Pricing

The CAPM is a major pricing method but it is by no means the only one. There are other alternatives; depending on the circumstances one may be more suitable than another. Risk-neutral pricing is just one such method. Unlike the CAPM, valuation of a security or project draws on other assets whose values or prices are already known. Below we make use of this method on several occasions.

2.3.1 State Prices

At the beginning of this chapter we have made it clear that, under uncertainty, a whole set of possible returns can arise from a single investment decision. Each of these outcomes corresponds to a given state (of nature). Thus, the number of potential states equals the number of possible outcomes. The decision maker obviously knows the prevailing state at the time of decision, but is typically unable to forecast the situation one period ahead. This inability to foresight perfectly the future state translates into uncertainty about the return on the investment. For this uncertainty to solve, the decision maker must wait one period at which time the state is revealed.

Let $j = 1, 2, \ldots, m$ denote the possible states of nature. We can think of a security ω as a vector of m components, $(\omega_1, \omega_2, \ldots, \omega_m)$, with ω_j describing its payoff in the jth state, i.e. if state j happens to occur. Of course, if there is a market for this security in equilibrium then it will command a price P. What is this price?

To answer this question we define a set of elementary securities s_j: they just pay one monetary unit in one precise state and zero otherwise (this is why they are sometimes referred to as state contingent claims); for example, s_3 entitles its holder to receiving the cash flows (0, 0, 1, 0, …, 0). Let P_j denote the price of the jth elementary security (provided it actually exists).

When there is one such elementary security for each possible state of nature the economy is said to have a market structure that is complete. In this case, the payoffs from any asset or security can be broken down into a weighted sum of payoffs from a collection of elementary securities:

$$(4, 2, 0, 0, 1) = 4s_1 + 2s_2 + 1s_5. \tag{2.20}$$

This in turn implies that the price of the former (P) will be just a weighted sum of those of the latter (P_j), with the weights/payoffs denoted by ω_j:

$$P = \sum_{j=1}^{j=m} \omega_j P_j. \tag{2.21}$$

It can also be the case that the $\{s_j\}$ do not exist as such but can be constructed synthetically from existing securities. Be it as it may, to the extent that they convey a positive payment in one state and zero otherwise their prices must be positive in equilibrium. Stated differently, if the latter were negative or zero there would be an arbitrage opportunity.

2.3.2 *Risk-Neutral Valuation*

We can normalize these (positive) state prices $\{P_j\}$ so as to make them sum to one. This trick is easily accomplished by dividing each P_j by the sum of all the $\{P_j\}$:

$$P_0 \equiv \sum_{j=1}^{j=m} P_j \rightarrow P_j \equiv \frac{P_j}{P_0} \rightarrow \sum_{j=1}^{j=m} P_j = 1 \tag{2.22}$$

This way we can interpret the resulting prices as probabilities.

We further observe in Eq. (2.21) that the $\{P_j = P_0 p_j\}$ multiply the payoffs $\{\omega_j\}$:

$$P = P_0 \sum_{j=1}^{j=m} \omega_j p_j. \tag{2.23}$$

Each term in the sum will thus be the product of a cash flow and its corresponding probability, i.e. an expected payoff. And since the $\{p_j\}$ add up to one, the whole sum will be the expected payoff from the security ω one period from now. At this stage it will only look natural to claim that today's price P is merely tomorrow's expected payoff discounted over one period:

$$P = P_0 \widehat{E}(\omega) = \frac{1}{(1 + rate)} \widehat{E}(\omega). \tag{2.24}$$

Here the expectation is not taken with respect to the actual probabilities of the possible states in the physical, real world (indeed, we have not even mentioned them). Instead, it is based on the $\{p_j\}$ that we have fabricated expressly; we denote it by $\widehat{E}$.

But, what is the appropriate discount rate? Note from Eq. (2.22) that P_0 is in fact the price of the riskless asset, which has a unitary payoff in all possible states:

$$P_0 \equiv \sum_{j=1}^{j=m} P_j = (1, 1, \cdots, 1, 1). \tag{2.25}$$

Given that this future payoff of 1 is sure, it must necessarily be discounted to the present at the risk-free rate r.

At the same time, r is also the discount rate used by investors who are risk neutral: they are indifferent between \$1 for sure and a lottery with expected payoff of \$1. Indifference in this case involves a common discount rate for the two \$1 payments, in particular, the riskless rate that applies naturally to risk-free assets. Thus:

$$P = \frac{1}{(1+r)} \widehat{E}(\omega). \tag{2.26}$$

We refer to this pricing method as risk-neutral valuation. Starting from the expectation tomorrow with respect to the risk-neutral probabilities $\{p_j\}$, the security price today is just this expectation discounted at the riskless rate.

Before leaving this section, note the following. The time -t futures price for delivery at T, denoted $F(T,t)$, is the value of the delivery price at time t such that the current (time t) value of the futures contract equals zero. It equals the expected spot price in a risk-neutral context. Thus, cash flows emanating from futures contracts can be properly discounted at r. Futures markets play a number of important roles. They convey information about the future trend in spot prices. They also reflect expectations about future demand and supply conditions. All this information is relevant for valuation purposes. It is also an input to decision making about production and/or storage. In addition, futures markets allow parties to hedge their positions in commodity inputs or outputs.

2.4 Forward and Futures Markets

2.4.1 A Primer

The price of raw materials or input fuels is a major issue when assessing the economic profitability of many industrial facilities. The same holds true for the price of the electricity output or that of other unintended byproducts such as polluting emissions (of sulfur, carbon, etc.). As it turns out, many of these commodities are customarily traded on futures markets. And futures prices are a key ingredient to the valuation process. Therefore, for those unfamiliar with the *raison d'etre* of these markets or the way in which they operate, the following introduction may be welcome. Along the way we also explain the relationship between futures prices and spot prices.

The standard explanation for the role of futures markets is that they help to spread and hence reduce risks, and to motivate the collection and dissemination of information relevant to the planning of consumption and production. Let us analyze these aspects in some detail.

A forward contract is an agreement whereby the seller currently agrees to deliver to the buyer a specific asset on a specified future date at a fixed price (K), to be paid on the delivery date (T). If the fixed price to be paid on the delivery date were sufficiently low, the buyer would have to pay a positive amount for the contract. If it were set high enough, the seller would have to pay the buyer to take the contract. Clearly, there is an intermediate price, known as the forward price, at which the current value of the contract would be zero (and both parties agree with). This is the fixed price that is customarily used for newly-written forward contracts. Consequently, a forward contract will have a value of zero when the contract is initiated (0). Of course, the value of an outstanding contract (*f*) will subsequently change as the value of the underlying asset changes.

One of the parties to a forward contract assumes a long position and agrees to buy the underlying asset on the delivery date for the delivery price. The other party assumes a short position and agrees to sell the asset on the same date for the same price. The forward price and the delivery price are equal at the time the contract is entered into. As time passes, the forward price is liable to change while the delivery price of that contract, of course, remains the same. A forward contract is settled at maturity. The holder of the short position delivers the asset to the holder of the long position in return for a cash amount equal to the delivery price (some futures markets allow settlement by differences in cash so there is no delivery of the physical asset).

Forward contracts are usually between two financial institutions or between a financial institution and one of its corporate clients. They are not normally traded on an exchange; they are traded over-the-counter (i.e. in OTC markets). No cash changes hand at the time of the agreement. However, either or both parties to the transaction often have to post some funds to guarantee fulfillment of the contract.

The easiest forward contract to value is one written on an asset that provides the holder with no income. Consider the following two portfolios at time t:

- Portfolio A: it consists of one long forward contract on the asset (worth f) plus an amount of cash equal to $Ke^{-r(T-t)}$ (or, equivalently, a zero-coupon bond that matures at the expiration date of the forward contract T and which at that date will have a value equal to the delivery price K).
- Portfolio B: it consists of one unit of the asset (worth S_t).

In Portfolio A, the cash, assuming that it is invested at the risk-free rate (r), will grow to an amount K at time T, to be used to pay for the security at the maturity of the forward contract. And the payoff from the long position in a forward contract on one unit of the asset will be $S_T - K$ at time T. Both portfolios will therefore consist of one unit of the asset at time T; see Table 2.1.

Table 2.1 Values of alternative portfolios

Portfolio	Current date	Delivery date
A: long forward	f	$S_T - K$
A: long cash	$Ke^{-r(T-t)}$	K
B: long asset	S_t	S_T

For there to be no arbitrage opportunities, they must be equally valuable at the earlier time t. It follows that, at time t, we have $f + Ke^{-r(T-t)} = S_t$. Obviously, this equation can be arranged into $S_t - f = Ke^{-r(T-t)}$, i.e., a portfolio consisting of one unit of the asset and one short forward contract would provide a certain amount of cash K at time T (regardless of whether the asset moves up or down in the meantime): $S_T - (S_T - K) = K$ (while of course $Ke^{-r(T-t)}e^{r(T-t)} = K$). Through the ability to construct such a riskless hedge, risk can be effectively "squeezed out" of the problem, so that investors' risk attitudes do not matter (Trigeorgis 1996). Therefore, for valuation purposes, we can equivalently -and more conveniently-pretend to be in a risk-neutral world where risk is irrelevant. In such a world, all assets (including stocks, options, forwards, futures, ...) would earn the risk-free return, and so expected cash flows (weighted by the risk-neutral probabilities) could be appropriately discounted at the risk-free rate.

Now similar to a forward contract, an individual who takes a long position in a futures contract nominally agrees to buy a designated good or asset on the delivery date (T) for the futures price prevailing at the time the contract is initiated (0) (thus, the party in a long position profits from a rise in the asset price). Again, no money changes hands initially. Subsequently, however, as the futures price changes, the party in whose favor the price change occurred must immediately be paid the full amount of the change by the losing party (each party's margin account thus fluctuates according to the change in the futures price). As a result, the payment required on the delivery date to buy the underlying good or asset is simply its spot price at that time. The difference between that amount and the initial futures price has been paid (or received) in installments throughout the life of the contract. Like the forward price, the equilibrium futures price must also continually change over time. It must do so in such a way that the remaining stream of future payments described above always has a value of zero.[1]

A futures contract may be defined as a highly standardized forward contract (Houthakker 1989). The standardization characteristic generally involves five elements: (1) Quantity: buyers and sellers can deal only in lots of fixed size (of course, they can buy or sell any number of such lots); (2) Quality: the commodity or asset is not usually completely specified, but can be anywhere in a range (e.g., all wheat of certain grades); (3) Delivery time: the lot can be delivered at any time within a specified period, say a month; (4) Location: the lot must be delivered in specified places in one or more specified cities; (5) Identity of contractors: after the initial contract is established, the buyer and seller normally have no further dealings with each other, thus eliminating credit risk. The execution is guaranteed by a clearing house, which acts as seller to all buyers and as buyer to all sellers. The clearing house can offer this guarantee by virtue of the security deposits, known as 'margin', it collects from its members.

---

[1] If interest rates are non-stochastic, there is no counterparty risk, and there are no arbitrage opportunities, it can be shown that futures prices are equal to forward prices.

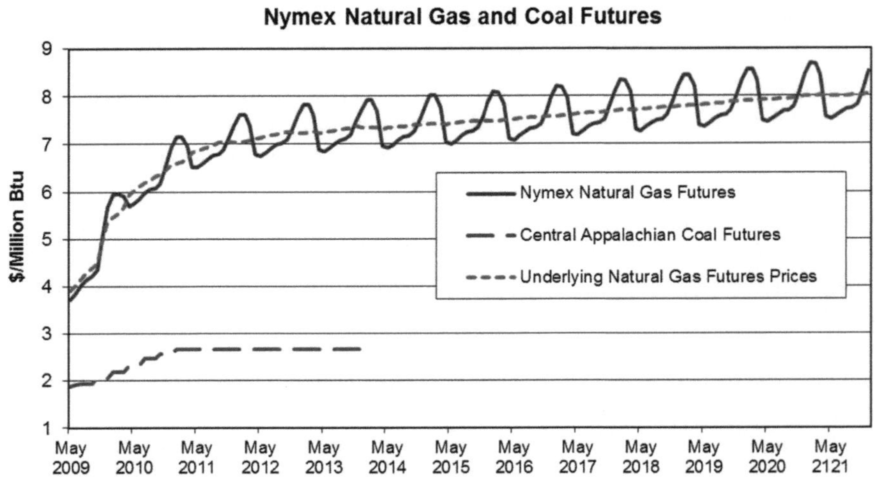

Fig. 2.4 Futures prices of coal and natural gas on NYMEX as of May 2009

The immediate purpose of this standardization is to minimize transaction costs and thereby to allow futures contracts to be traded on an organized exchange. Futures prices are regularly reported in the financial press. They are determined on the floor of the exchange in the same way as other prices (i.e., by the laws of supply and demand). For example, a risk-averse wheat farmer may hedge his future harvest by selling October wheat futures in January, in which case he is 'long' in actuals and 'short' in futures. A trader whose net position in the spot market is offset by his position in the futures market is called a hedger; in particular, he is a 'short hedger' if he is long in the spot market and short in the futures market (and a 'long hedger' if these positions are reversed). Traders who are net long or net short in the overall market (spot plus futures) are known as speculators. Figure 2.4 displays futures coal and natural gas prices on the New York Mercantile Exchange (NYMEX) as of May 2009 for a number of maturity dates. Natural gas prices in particular show a clear seasonal pattern, which can be subtracted from the original time series. Contract maturities are much longer for natural gas than for coal.

The prices prevailing in the spot and futures markets at any time are not necessarily equal. However, there are two main links between these markets; one is provided by the delivery mechanism and the other by hedging. As to delivery, when a futures contract reaches maturity the remaining shorts have to deliver what they have sold, and the remaining longs have to accept and pay for what they have bought. Clearly, the shorts will not deliver anything that could be sold at a higher price in the spot market, nor will the longs take delivery of anything that they could buy more cheaply elsewhere. At delivery time, therefore, the futures price must be equal to the spot price of the items that are actually delivered. Since this ultimate equality is widely anticipated, it will also influence futures and spot prices prior to delivery time.

Hedging also serves to relate futures prices and spot prices. If a futures price is high compared to a spot price, hedgers will buy in the spot market and sell futures. They can do so without risk if the futures price exceeds the spot price by more than the carrying charge, which is the cost of holding physical inventories between the present and the maturity of the futures contract. The futures price therefore cannot exceed the current spot price by more than the prevailing carrying charge.[2]

If merchants can increase their profits by hedging, they must be willing to pay a risk premium for the opportunity to do so. It is conceivable that short hedging exactly offsets long hedging, in which case any premiums paid by hedgers would cancel out. There is considerable evidence, however, that in most markets short hedging exceeds long hedging at most times. Now if the hedgers are net short in futures, the speculators in futures must be net long (since there is a sale for every purchase). It can be argued that speculators will only be net long if they expect futures prices to rise. At any particular moment the speculators may of course be wrong, but on average they are right, and each futures price will tend to rise until, at the maturity of the contract, it equals the relevant spot price. The speculators' gain is the hedgers' loss; thus the speculators receive a risk premium proportionate to the amount of hedging they make possible. This risk premium is implicit in the hedgers' willingness to sell futures contracts that have a tendency to appreciate: the futures price is below the expected future spot price. This, in brief, is the so-called theory of normal backwardation.[3]

It is also possible at times that hedgers are net long in futures while speculators are net short. Then the opposite pattern would result: the futures price is above the expected future spot price, so futures prices will tend to fall. This situation is called contango. To be sure, the price of a futures contract will always reflect the combination of two elements: (1) the traders' forecast of the future spot price of the asset, and (2) the traders' risk aversion (in the form of the implicit risk premium for securing a fixed price today for future delivery).

In a risk-neutral context, all the individuals are assumed to be risk neutral. As such, they are only concerned with average or expected values (not with dispersion around these values): as long as all the available assets (or portfolio of assets) have the same expected returns, the individuals are indifferent between them. One particular asset is the risk-free asset (e.g., a Treasury bill, assuming its probability of default is null). Let r denote the (certain) rate of return on this asset. If all the individuals show risk neutrality, all the assets in the market must have the same

---

[2] It does not follow, however, that a futures price must always exceed the spot price by the relevant carrying charge. Positive inventories may be held even if the spot price is above the futures price. This is because inventories have a 'convenience yield convenience' derived from their availability when buyers need them.

[3] The empirical validity of this theory remains in dispute.

expected rate of return, in particular the risk-free rate r.[4] Further, it can be shown that the futures price $F(T, t)$, i.e. the value of the delivery price at time t such that the current value of the futures contract equals zero, is the expected spot price in a risk-neutral context. In this context, expected cash flows from any one asset can be discounted to the present at the riskless rate r in order to compute their present value. Of course, this applies to cash flows from futures contracts since they are risk free (the clearinghouse guarantees the performance of the parties to each transaction).

2.4.2 Futures Prices, Spot Prices, and Storage Costs

As a starting point, consider a futures contract which is written on a security that provides the holder with no income. If the contract is to have zero value at inception and there are to be no arbitrage opportunities, the relationship between the futures price F and the spot price S for a no-income security must be:

$$F = S_t e^{r(T-t)}. \tag{2.27}$$

Now assume that the underlying security provides a perfectly predictable dividend yield which is paid continuously at a yearly rate y. In this case we get:

$$F = S_t e^{(r-y)(T-t)}. \tag{2.28}$$

Consider, instead, that the underlying asset is a commodity held mainly for investment purposes (e.g. gold, silver). The storage costs (here assumed proportional to commodity price S) can be interpreted as a negative dividend yield being paid at a rate u (see Hull (2005)):

$$F = S_t e^{(r+u)(T-t)}. \tag{2.29}$$

A number of commodities, however, are held in inventory by firms and individuals because of their consumption value; this is in stark contrast to futures contracts, which cannot be consumed. The benefits from owning the physical commodity can be so high that businesses forego seemingly arbitrage opportunities. Thus, they can decide to cling to their corn bushels instead of selling them and buying corn futures despite the observation that $S_t e^{(r+u)(T-t)} > F$. These benefits are usually referred to as the convenience yield δ, which is implicitly defined by:

$$Fe^{\delta(T-t)} = S_t e^{(r+u)(T-t)} \rightarrow F = S_t e^{(r+u-\delta)(T-t)}. \tag{2.30}$$

[4] If one particular asset had a higher expected return, then all the agents would chase it. But every purchase requires that there is someone willing to sell, which is not the case. The same holds true if a particular asset has an expected rate of return lower than r: everyone tries to sell it but nobody is willing to buy it. Market equilibrium in a risk-neutral setting requires that the expected return is just the same across all the assets.

If the market anticipates shortages prior to the delivery date, the convenience yield will be relatively high. Conversely, if users' inventories of the commodity run high, then shortages will loom small and δ will be relatively low.

In the next chapter we will introduce several stochastic processes that the price of the underlying asset S can follow. The presence of price risk will almost inevitably call for a risk premium in the pricing equation. Assume, for example, that S evolves according to a random walk. Let $\alpha \equiv \frac{1}{S}\frac{E(S)}{dt}$ and λ denote the instantaneous growth rate of S in the physical world and the risk premium (assumed proportional to S), respectively. As shown in the next chapter, the futures price for delivery at T is related to the spot price S_t (through the non-arbitrage principle) by:

$$F = S_t e^{(\alpha-\lambda)(T-t)}. \tag{2.31}$$

At that point the reader can perhaps feel that there is a discontinuity with the economics of storage. Yet this is by no means the case, as we explain in Chap. 3.

As it turns out, the difference $(\alpha - \lambda)$ is equivalent to the difference $(r - \delta)$. Therefore, the futures price can be equivalently determined by:

$$F = S_t e^{(\alpha-\lambda)(T-t)} = S_t e^{(r-\delta)(T-t)}. \tag{2.32}$$

As explained above, if there is a cost to storing the commodity, and the storage cost per unit time is a percent of the spot price, then Eq. (2.32) changes into:

$$F = S_t e^{(r+u-\delta)(T-t)}. \tag{2.33}$$

In order to undertake valuations based on futures prices, we will need the joint parameter in parentheses, be it approximated either through $(\alpha - \lambda)$ or alternatively by $(r + u - \delta)$; we do not need any of the individual parameters in isolation. Thus, storage costs in particular are already included in the parameter estimates; the model behaves the same irrespective of whether these costs are high or low. In the particular case of carbon emission allowances, it seems reasonable to assume that u would be zero or close to zero. The relative size for other commodities (e.g. coal) is also minor. For example, storage costs for natural gas in the U.S. have been in the range [0.0006, 0.0029] $/thousand cubic feet (TCF) according to FERC (2004), while gas price has ranged between 3.9 and 8.1 $/TCF from 2000 to 2010 (EIA 2011). Again, knowledge of its particular value is irrelevant for the valuations undertaken here (as long as the net composite is estimated correctly).

References

Elton EI, Gruber MJ, Brown SJ, Goetzmann WN (2009) Modern portfolio theory and investment analysis, 8th edn. Wiley, New York

Houthakker HS (1989) Futures Trading. In: Eatwell J, Milgate M, Newman P (eds) The new Palgrave: finance. The Macmillan Press Ltd., New York, pp 153–158

Hull J (2005) Options, Futures, and Other Derivatives, 6th edition. Prentice Hall, Englewood Cliffs, New Jersey

Luenberger DG (2009) Investment science. Oxford University Press, New York

Markowitz HM (1987a) Mean-variance analysis in portfolio choice and capital markets. Basil Blackwell, New York

Markowitz HM (1987b) Portfolio selection. Wiley, New York

Sharpe WM (1964) Capital asset prices: a theory of market equilibrium under conditions of risk. J Finan 19(3):425–442

Trigeorgis L (1996) Real options—managerial flexibility and strategy in resource allocation. The MIT Press, Cambridge, MA

US Energy Information Administration (2011) http://www.eia.gov/dnav/ng/hist/n9190us3A.htm

US Federal Energy Regulatory Commission (2004) Current state of and issues concerning underground natural gas storage. Staff report, Sep 30

Chapter 3
Analytical Solutions

3.1 Stochastic Price Models

It is hardly a surprise that market prices can go up and down. This holds true whether we are dealing with financial assets (say, a common stock) or real assets (a piece of land). As a consequence, the value of any derivative asset (like an option, whose return depends on that of the underlying asset, say, a barrel of oil) becomes also uncertain. Derivative pricing models allow us to determine the price of a derivative asset as a function of a set of observable variables. No doubt this feature has been a main driver for the success of these tools to the investment profession.

Research on the behavior of commodity prices has been intense for decades. Mean reversion has been frequently observed in a number of commodity prices. Yet there is hardly a universal consensus on the stochastic process that best fits the behavior of commodity prices. This subject will probably keep on going debated for some time.

The starting point might be stated following Dixit and Pindyck (1994): "Are the prices of raw commodities and other goods best modeled as geometric Brownian motions or as mean-reverting processes? One way to answer this is to examine the data for the price variable in question… It usually requires many years of data to determine with any degree of confidence whether a variable is indeed mean-reverting". A similar argument can be found in Baker et al. (1998): "Econometricians have tests for stationarity. One group of tests, including the Dickey-Fuller test, focus on finding a 'unit root' in the time-series of the commodity price. Unfortunately these tests have fairly low power and unless a large number of observations are available over a long time period it is difficult to reject the null hypothesis of a random walk even when the series is generated by a reverting process". As an illustration, Dixit and Pindyck (1994) analyze the case of crude oil and copper prices, in constant 1967 dollars, over 120 years: "Running these tests on the full 120 years of data, one can easily reject the random walk hypothesis; that is, the data confirm that the prices are mean reverting. However, if one performs unit root tests using data for only the past 30 or 40 years, one fails to reject the random walk hypothesis. This seems to be the case for many other economic variables as well".

L.M. Abadie and J.M. Chamorro, *Investment in Energy Assets Under Uncertainty*, Lecture Notes in Energy 21, DOI: 10.1007/978-1-4471-5592-8_3,

Faced with this weakness they conclude: "As a result, one must often rely on theoretical considerations (for example, intuition concerning the operation of equilibrating mechanisms) more than statistical tests when deciding whether or not to model a price or another variable as a mean-reverting process".

There seems to be, nonetheless, further sources of information. According to Baker et al. (1998): "Many commodities have traded futures or forward contracts, and the price series for these contracts are additional sources of information about the dynamics of the underlying spot price of the commodity. Even when data on spot prices does not provide clear evidence of reversion, data on futures prices often strongly supports the hypothesis that there is reversion in commodity prices". In particular, they claim that "All that matters for many valuations is the expected spot price under the risk-adjusted distribution. Futures prices enable us to estimate the parameters of the risk-adjusted distribution, even when they do not significantly improve our estimates of all of the parameters of the underlying distribution or process driving the commodity price". They provide further evidence from the term structure of futures prices: "The [higher] volatility in the spot price reflects temporary shocks to supply and demand. Since these shocks dissipate over time, the futures price for longer maturity contracts reflect only the small residual effect that is expected to remain by the time of maturity. Consequently the futures price series is less volatile than the spot price series. This lower volatility for the longer maturity futures contracts is an important fact suggesting that the oil price is reverting and cannot be accurately described as a random walk". Cortazar and Schwartz (2003) share their view: "[The] random walk specification for commodity prices was used until a decade ago, when mean reversion in spot prices began to be included as a response to the evidence that volatility of futures returns declines with maturity".

As a consequence, Baker et al. (1998) claim that: "Most commodity prices do not follow a random walk, and accurate valuations often require that the analyst go beyond this familiar model". Similar conclusions appear in Ronn (2002): "Empirically, spot prices for natural gas, electricity, industrial metals, and other commodities display mean reversion"; and Pilipovic (1998): "As will be seen from our analysis results, energy markets require mean-reverting models. In fact, the price mean-reverting model turns out to do the best job of capturing the distribution of energy prices". Recent contributions to this literature where mean reversion appears prominently are Benth et al. (2012), Keles et al. (2012), and Nomikos and Andriosopoulos (2012).

3.1.1 The Geometric Brownian Motion

A number of asset prices are traditionally modeled as a Geometric Brownian Motion (GBM), e.g., the behavior of a common stock:

$$dS_t = \alpha S_t dt + \sigma S_t dZ_t,$$

where S_t denotes the time-t (spot) price of the stock (or, for example, an allowance to emit 1 metric ton of CO_2). As usual, α stands for the expected growth rate, and σ for the instantaneous volatility of (carbon) price changes. dZ_t is the increment to a standard Wiener process; it is assumed to be normally distributed with zero mean and variance dt. Price returns dS_t/S_t are thus assumed to behave normally. But $dS_t/S_t = d(\ln S_t)$; therefore, the price S_t follows a lognormal distribution.

Stochastic differential equations (SDEs) typically result from incorporating random fluctuations in the system's dynamic description, e.g., a Gaussian white noise. They are written symbolically as stochastic differentials, for example the GBM above. Nonetheless, they are interpreted as integral equations with stochastic integrals; in the case of the GBM:

$$S_t = S_{t_0} + \int_{t_0}^{t} \alpha S_s \, ds + \int_{t_0}^{t} \sigma S_s \, dZ_s. \tag{3.1}$$

The second integral is an Ito integral which involves the unknown solution. In this particular case we can find an explicit formula for the solution of the SDE:

$$S_t = S_0 e^{\left(\alpha - \frac{1}{2}\sigma^2\right)t + \sigma Z_t}. \tag{3.2}$$

But this will not be the case in general; we must then resort to some numerical methods to determine the solution approximately. Figure 3.1 displays a number of simulated GBMs; they assume a starting (deseasonalised) carbon price of 10 $/t$CO_2$ with a (risk-neutral) drift rate of 5 %, and volatility 40 %.

Now, adopting the transformation $X_t \equiv \ln S_t$ and applying Ito's Lemma yields:

$$dX_t = \left(\alpha - \frac{\sigma^2}{2}\right) dt + \sigma dZ_t. \tag{3.3}$$

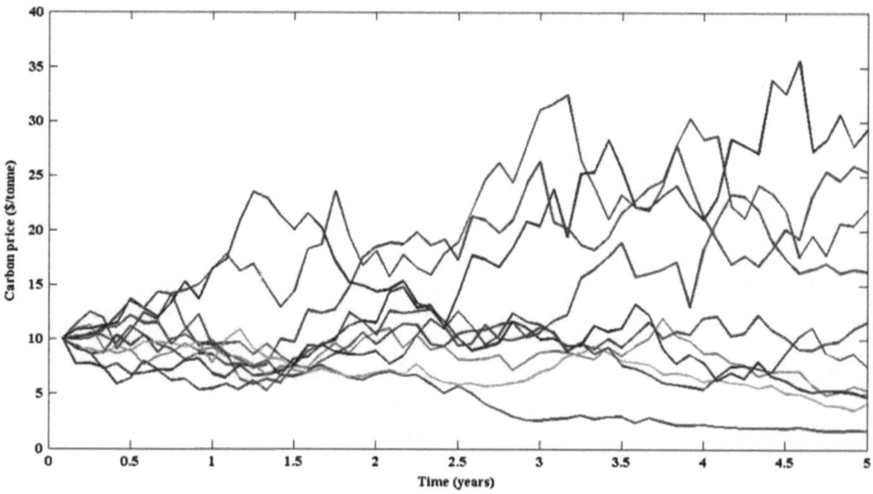

Fig. 3.1 Simulated random paths following a GBM

We will use this specification a number of times in the next chapters.

Box 3.1 Stochastic calculus and Ito's Lemma

The value of a stock option is a function of the stock price. This in turn is usually modelled as a stochastic (Ito) process. In order to differentiate or integrate functions of such processes we need a tool known as Ito's Lemma.

Consider an asset whose price dynamics is governed by:

$$dS = a(S,t)dt + b(S,t)dZ,$$

with $a(\cdot)$ and $b(\cdot)$ nonrandom. Let $F(S,t)$ be a function of the underlying asset and possibly calendar time (e.g., the value of an option); assume that it is at least twice differentiable in S and once in t. How does F behave, i.e., what is dF? Ito's Lemma provides the answer:

$$dF(S,t) = \frac{\partial F}{\partial t}dt + \frac{\partial F}{\partial S}dS + \frac{1}{2}\frac{\partial^2 F}{\partial S^2}(dS)^2.$$

In expanded form we have:

$$\begin{aligned} dF(S,t) =& \frac{\partial F}{\partial t}dt + \frac{\partial F}{\partial S}[a(S,t)dt + b(S,t)dZ] + \frac{1}{2}\frac{\partial^2 F}{\partial S^2}\left[b^2(S,t)dt\right] \\ =& \left[\frac{\partial F}{\partial t} + a(S,t)\frac{\partial F}{\partial S} + \frac{1}{2}b^2(S,t)\frac{\partial^2 F}{\partial S^2}\right]dt + b(S,t)\frac{\partial F}{\partial S}dZ. \end{aligned}$$

It is possible to extend Ito's Lemma to functions of several processes.

Moments In terms of the notation in Box 3.2, the GBM corresponds to the case $a_1(t) = \alpha$ and $b_1(t) = \sigma$ with $a_2(t) = b_2(t) = 0$. Consider the first moment $m(t) \equiv E(S_t)$; following Kloeden and Platen (1992) we have:

$$\frac{dE(S_t)}{dt} = \alpha E(S_t).$$

Using an integration factor $e^{-\alpha t}$ we get:

$$\frac{dE(S_t)}{dt}e^{-\alpha t} - \alpha E(S_t)e^{-\alpha t} = 0.$$

Integrating this expression yields:

$$E(S_t)e^{-\alpha t} = c,$$

where $c = S_0$. Hence:

$$E(S_t) = S_0 e^{\alpha t}. \tag{3.4}$$

Regarding the variance of S_t again we follow Kloeden and Platen (1992). The second order moment $P(t) \equiv E(S_t^2)$ must satisfy:

$$\frac{dP(t)}{dt} = \left(2\alpha + \sigma^2\right)P(t).$$

Using the integration factor $e^{-(2\alpha+\sigma^2)t}$ yields:

$$\frac{dP(t)}{dt}e^{-(2\alpha+\sigma^2)t} - \left(2\alpha + \sigma^2\right)P(t)e^{-(2\alpha+\sigma^2)t} = 0.$$

Integrating this differential equation:

$$P(t)e^{-\left(2\alpha+\sigma^2\right)t} = c,$$

with the constant of integration $c = S_0^2$. Therefore,

$$P(t) = S_0^2 e^{\left(2\alpha+\sigma^2\right)t}.$$

To get the formula for the variance of S_t we substitute this second moment and the first moment squared:

$$Var(S_t) = S_0^2 e^{\left(2\alpha+\sigma^2\right)t} - S_0^2 e^{2\alpha t} = S_0^2 e^{2\alpha t}\left(e^{\sigma^2 t} - 1\right). \tag{3.5}$$

Here the variance increases with time without bound: $\mathbf{lim}_{t\to\infty} \boldsymbol{Var}(\boldsymbol{S_t}) \to \infty$.

Box 3.2 Stochastic differential equations and their solution
Consider the following linear SDE:

$$dS_t = [a_1(t)S_t + a_2(t)]dt + [b_1(t)S_t + b_2(t)]dZ_t,$$

where a_1, a_2, b_1, and b_2 are known functions of time t or constants. Now we compute the first two moments of this process. According to Kloeden and Platen (1992), denoting $m(t) \equiv E(S_t)$ and $P(t) \equiv E(S_t^2)$, we have:

$$\frac{dm(t)}{dt} = a_1(t)m(t) + a_2(t),$$

$$\frac{dP(t)}{dt} = \left[2a_1(t) + b_1^2(t)\right]P(t) + 2m(t)[a_2(t) + b_1(t)b_2(t)] + b_2^2(t).$$

These two equations are linear and can be solved using integrating factors.

Risk-neutral version In order to obtain the risk-neutral version of the GBM process, we subtract a risk premium to its actual rate of growth. Assuming that the former is proportional to the asset price S, and according to the CAPM, it is $\rho\sigma\phi S$. In this expression, ρ is the correlation between the returns on the market portfolio and the commodity asset. ϕ denotes the market price of risk, which is defined as $\phi \equiv \frac{r_M - r}{\sigma_M}$, where r_M is the expected return on the market portfolio. And σ_M denotes its volatility. For notational simplicity, $\rho\sigma\phi \equiv \lambda$. Thus:

$$\frac{dS}{S} = (\alpha - \lambda)dt + \sigma dZ. \tag{3.6}$$

Following the same steps as before it can be shown that:

$$E^Q(S_t) = S_0 e^{(\alpha - \lambda)t}; \tag{3.7}$$

here the superscript Q denotes the mathematical expectation under risk neutrality (or "the equivalent martingale measure", as it is also known).

We have $\alpha = (1/S)(E(S)/dt)$. Assume there is a rate of return shortfall (or the so-called convenience yield) δ, which is a rather usual concept in the case of energy commodities; it reflects the profits enjoyed by the owner of the physical commodity, as opposed to the holder of a futures contract (it is equivalent to the dividends received by the holder of a firm's stock, as opposed to the holder of a stock option). Thus, $\mu = \alpha + \delta$ denotes the total expected rate of return. Now "this expected return must be enough to compensate the holders for risk. Of course it is not risk as such that matters, but only non-diversifiable risk. The whole market portfolio provides the maximum available diversification, so it is the covariance of the rate of return on the asset with that on the whole market portfolio that determines the risk premium. The fundamental condition of equilibrium from the

CAPM says that $\mu = r + \phi\sigma\rho$" (Dixit and Pindyck 1994, p. 115). Hence, $\alpha + \delta = \mu = r + \phi\sigma\rho$. This implies that $\alpha - \phi\sigma\rho = r - \delta$. Therefore, we can use interchangeably either $(\alpha - \lambda)$ or $(r - \delta)$ in the valuation formulas below; i.e., they can be equivalently stated in terms of the actual growth rate minus the risk premium, or the riskless interest rate minus the rate of return shortfall (or convenience yield).

Again, defining $X_t \equiv \ln S_t$ and applying Ito's Lemma the risk-neutral version happens to be:

$$dX_t = \left(\alpha - \frac{\sigma^2}{2} - \lambda\right) dt + \sigma dZ_t. \tag{3.8}$$

We will use this specification frequently in the next chapters.

3.1.2 The Inhomogenous Geometric Brownian Motion

In a model for long-term valuation of energy assets, it is convenient to keep in mind that prices tend to revert toward levels of equilibrium after an incidental change. From the models which display mean reversion we have chosen the Inhomogeneous Geometric Brownian Motion (or IGBM) process:

$$dS_t = k(S_m - S_t)dt + \sigma S_t dZ_t, \tag{3.9}$$

where S_t denotes the price at time t while S_m stands for the level which the price tends to in the long run. The parameter k is the speed of reversion toward the "normal" level. It can be computed as $k = \ln 2/t_{1/2}$, where $t_{1/2}$ is the expected half-life, that is the time for the gap between S_t and S_m to halve. σ is the instantaneous volatility of price, which determines the variance of S_t at t. And dZ_t denotes the increment to a standard Wiener process; it is normally distributed with mean zero and variance dt. Figure 3.2 displays a number of simulated IGBMs; they assume a starting coal price of 90 \$/t with a risk-neutral long-term level of 140 \$/t, and volatility 20 %.

Some of the reasons for our choice are: (a) This model satisfies the following condition (which seems reasonable): if the price of one unit of the commodity reverts to some mean value, then the price of two units reverts to twice that same mean value. (b) The term $\sigma S_t dZ_t$ in the differential equation precludes, almost surely, the possibility of negative values. (c) The GBM is nested in this model, i.e., the latter admits $dS_t = \alpha S_t dt + \sigma S_t dZ_t$ when $S_m = 0$ and $\alpha = -k$ as a particular case. (d) The expected value in the long run is: $E(S_\infty) = S_m$; this is not true in Schwartz (1997 Model 1), where $E(S_\infty) = S_m e^{-\sigma^2/4k}$.

Statistical moments Leaving aside any seasonal behavior, in general we have:

$$dS_t = k(S_m - S_t)dt + \sigma S_t dZ_t.$$

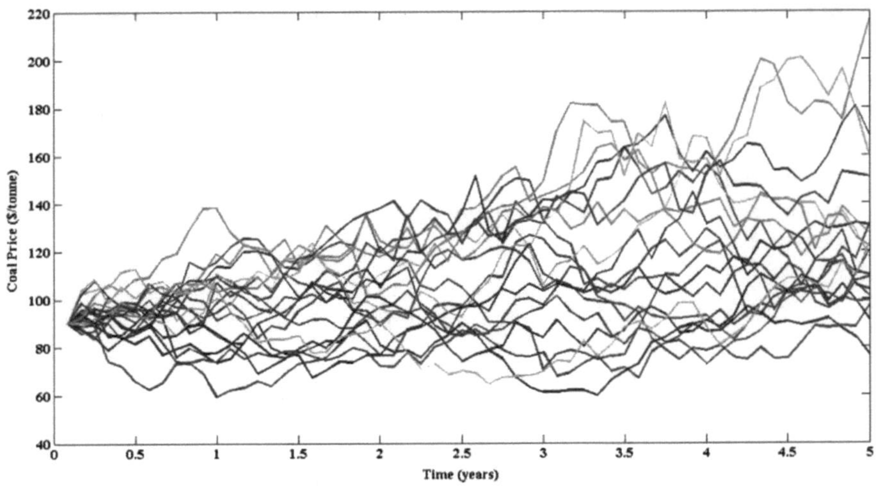

Fig. 3.2 Simulated random paths following an IGBM

In terms of the notation in Box 3.2, the IGBM corresponds to the case: $a_1 = -k$, $a_2 = kS_m$, $b_1 = \sigma$ and $b_2 = 0$. In this case, the expected value satisfies the following differential equation (Kloeden and Platen 1992):

$$\frac{dE(S_t)}{dt} = -kE(S_t) + kS_m.$$

Rearranging and using an integration factor e^{kt} we get:

$$\frac{dE(S_t)}{dt} e^{kt} + kE(S_t)e^{kt} = kS_m e^{kt}.$$

Integrating yields:

$$E(S_t)e^{kt} = S_m e^{kt} + c;$$

for $t = 0$ the constant equals $c = S_0 - S_m$. Therefore:

$$E(S_t) = S_m + (S_0 - S_m)e^{-kt} = S_m\left(1 - e^{-kt}\right) + S_0 e^{-kt}. \tag{3.10}$$

Hence we can see that:

$$\lim_{k\to\infty} E(S_t) = S_m, \quad \lim_{t\to\infty} E(S_t) = S_m.$$

It can be verified easily that when $\boldsymbol{k} = -\boldsymbol{\alpha}$ and $\boldsymbol{S_m} = 0$, the expected value of the GBM model results: $\boldsymbol{E(S_t) = S_0 e^{\alpha t}}$. When there is seasonality then it suffices to add the seasonal behavior $\boldsymbol{f(t)}$ to the (deseasonalized) expected value.

There is a relationship between k and the time $t_{1/2}$ at which the expected value equals the mid point between S_0 and S_m:

$$S_m + (S_0 - S_m)e^{-kt_{1/2}} = \frac{S_0 + S_m}{2}.$$

From this expression it follows that:

$$t_{1/2} = \frac{\ln 2}{k}.$$

For the second order moment of an IGBM process, the ordinary differential equation is (Kloeden and Platen 1992):

$$\frac{dP(t)}{dt} = \left(-2k + \sigma^2\right)P(t) + 2E(S_t)kS_m.$$

After substituting and rearranging, this can be rewritten as:

$$\frac{dP(t)}{dt} + \left(2k - \sigma^2\right)P(t) = 2kS_m\left[S_m + (S_0 - S_m)e^{-kt}\right].$$

Using an integration factor $e^{\left(2k-\sigma^2\right)t}$:

$$e^{\left(2k-\sigma^2\right)t}P(t) = 2kS_m \int_0^t e^{\left(2k-\sigma^2\right)t}\left[S_m + (S_0 - S_m)e^{-kt}\right]dt + c.$$

After some algebra:

$$P(t) = \frac{2kS_m^2}{2k - \sigma^2}\left[1 - e^{\left(\sigma^2-2k\right)t}\right] + \frac{2kS_m(S_0 - S_m)}{k - \sigma^2}\left[e^{-kt} - e^{\left(\sigma^2-2k\right)t}\right] + S_0^2 e^{\left(\sigma^2-2k\right)t},$$

where we have substituted S_0^2 for the constant c so that at $t = 0$ the moment takes on the value S_0^2. Now, from this formula we can derive the explicit solution for the variance:

$$\begin{aligned} Var(S_t) = E\left[(S_t - E(S_t))^2\right] &= e^{\left(\sigma^2-2k\right)t}\left[S_0^2 + \frac{2kS_m^2}{\sigma^2 - 2k} + \frac{2kS_m(S_0 - S_m)}{\sigma^2 - 2k}\right] \\ &+ e^{-kt}\left[\frac{2kS_m(S_0 - S_m)}{k - \sigma^2} + 2S_m(S_m - S_0)\right] - e^{-2kt}(S_0 - S_m)^2 + \frac{2kS_m^2}{2k - \sigma^2} - S_m^2. \end{aligned} \tag{3.11}$$

Thus, if we want to value an asset that conforms to this model, for high values of k there is no risk. In this case, for high values of k, cash flows should be discounted at the risk-free rate r. Therefore:

$$\lim_{k\to\infty} Var(S_t) = 0.$$

The variance of $\boldsymbol{S_t}$ would be almost zero when k is high, despite the existence of a certain volatility $\boldsymbol{\sigma}$ which in the short-term can push the value $\boldsymbol{S_t}$ to levels that are far from the equilibrium value $\boldsymbol{S_m}$. Mean reversion, which is a rather usual behavior in commodity prices, dampens the volatility of future cash flows as compared to a GBM model. This in turn implies a lower discount rate, in particular when discounting cash flows arising in the long run. Failure to consider this behavior can lead us to valuing wrongly long-term investments, such as those in energy assets with decades-long useful lives.

Depending on the value of k in relation to σ^2, this model implies:

$$for\ \frac{\sigma^2}{2} \geq k : \frac{dVar(S_t)}{dt} > 0 \rightarrow \lim_{t\rightarrow\infty} Var(S_t) \rightarrow \infty;$$

$$for\ \frac{\sigma^2}{2} < k : \lim_{t\rightarrow\infty} Var(S_t) = \frac{2kS_m^2}{2k - \sigma^2} - S_m^2;\ if k \rightarrow \infty : \lim_{t\rightarrow\infty} Var(S_t) = 0.$$

Thus, the relationship between k and $\boldsymbol{\sigma}^2$ determines the level of risk. Even though the prices on a given market show a great volatility, a strong reversion to the mean can imply a low level of real risk.

In subsequent sections it will be convenient to use simpler expressions for the numerical computations. It is known that when the increment Δt is very small, given $e^{at} = 1 + at + \frac{(at)^2}{2} + \cdots$, then $e^{a\Delta t} \approx 1 + a\Delta t$. Substituting in the equations for $E(S_t)$ and $E(S_t^2)$ the usual results of Euler-Maruyama's approximation arise:

$$E(S_t) \approx S_{t-1} + k(S_m - S_{t-1})\Delta t,$$

$$Var(S_t) \approx S_{t-1}^2 \sigma^2 \Delta t.$$

Finally, when $\boldsymbol{S_m} = 0$ and $\boldsymbol{k} = -\boldsymbol{\alpha}$ (i.e., the GBM) we get the expected value $\boldsymbol{S_0 e^{\alpha t}}$, the second order moment $\boldsymbol{S_0^2 e^{(\sigma^2+2\alpha)t}}$, and the variance $\boldsymbol{S_0^2 e^{(\sigma^2+2\alpha)t} - S_0^2 e^{2\alpha t} = S_0^2 e^{2\alpha t}} \left[\boldsymbol{e^{\sigma^2 t}} - 1\right]$. In sum, we get the formulas corresponding to a standard GBM process.

Risk-neutral version For our valuation purposes below we will follow the risk-neutral valuation approach. The change from an actual process to a risk-neutral one is accomplished by subtracting the appropriate risk premium ($\boldsymbol{\lambda} \equiv \boldsymbol{\phi\sigma\rho}$) from the actual growth rate (in the GBM case, $\boldsymbol{\alpha}$).[1]

Now let S_t denote the risk-neutral version; thus, without seasonality:

$$dS_t = [kS_m - (k + \lambda)S_t]dt + \sigma S_t dZ_t. \tag{3.12}$$

Following Kloeden and Platen (1992), the first moment must satisfy:

$$\frac{dE^Q(S_t)}{dt} = -(k + \lambda)E^Q(S_t) + kS_m.$$

[1] This is equivalent to replacing the drift in the price process (in the GBM case, α) with the growth rate in a risk-neutral world, $r - \delta$, where r is the riskless interest rate and δ denotes the net convenience yield. Note, though, that the convenience yield is not constant in a mean-reverting process.

Rearranging and using an integration factor $e^{(k+\lambda)t}$ we get:

$$\frac{dE^Q(S_t)}{dt}e^{(k+\lambda)t} + (k+\lambda)E^Q(S_t)e^{(k+\lambda)t} = kS_m e^{(k+\lambda)t}.$$

When $t = 0$ it must be $S_0 = c$. Therefore:

$$E^Q(S_t) = \frac{kS_m}{k+\lambda}\left(1 - e^{-(k+\lambda)t}\right) + S_0 e^{-(k+\lambda)t}. \tag{3.13}$$

If $t \to \infty$ then $E^Q(S_t)$ approaches $kS_m/(k+\lambda)$ in the long term under risk neutrality.

3.2 Annuities and Futures Contracts Under the Above Processes

3.2.1 Annuities Under the GBM

We start from the stochastic differential equation under risk neutrality:

$$dS_t = (\alpha - \lambda)S_t dt + \sigma S_t dZ_t.$$

Now we take the mathematical expectation:

$$E(dS_t) = (\alpha - \lambda)S_t dt.$$

Integrating this expression yields:

$$E(S_t) = S_0 e^{(\alpha-\lambda)t}.$$

To derive the value of an annuity between times τ_1 and τ_2 requires computing the following integral:

$$PV = \int_{\tau_1}^{\tau_2} E(S_t)e^{-rt}dt = \int_{\tau_1}^{\tau_2} S_0 e^{(\alpha-\lambda)t}e^{-rt}dt.$$

The solution is:

$$PV = \frac{S_0}{\alpha - \lambda - r}\left[e^{(\alpha-\lambda-r)\tau_2} - e^{(\alpha-\lambda-r)\tau_1}\right]. \tag{3.14}$$

Box 3.3. Example: One ton CO_2 per year avoided over 20 years. Assume that the carbon allowance price follows a GBM process. The initial price is 10 euros per ton CO_2, the risk-neutral drift rate is 5 %, and the riskless rate is 2 %. An annuity received continuously over 20 years is worth:

$$PV = \frac{10}{0.05 - 0.02}\left[e^{0.03\times 20} - 1\right] = 274.04\,\text{euros}.$$

This amount can be compared to the initial investment required to avoid the ton of CO_2. If there were a delay of τ_1 periods (say, 1 year) between the initial disbursement and the first reductions the savings would amount to:

$$PV = \frac{10}{0.05 - 0.02}\left[e^{0.03\times 21} - e^{0.03\times 1}\right] = 282.39\,\text{euros}.$$

3.2.2 Annuities Under the IGBM

We start start from the stochastic differential equation under risk neutrality:

$$dS_t = [kS_m - (k + \lambda)S_t]dt + \sigma S_t dZ_t,$$

where $\lambda \equiv \rho\sigma\phi$. The GBM is nested in this model for $S_m = 0$ and $\alpha = -k$. We know that:

$$E^Q(S_t) = \frac{kS_m}{k+\lambda}\left(1 - e^{-(k+\lambda)t}\right) + S_0 e^{-(k+\lambda)t}.$$

The value of an annuity between times τ_1 and τ_2 requires computing the integral:

$$PV = \int_{\tau_1}^{\tau_2} E^Q(S_t)e^{-rt}dt,$$

which yields:

$$PV = \frac{S_0 - \frac{kS_m}{\lambda+k}}{\lambda + k + r}\left[e^{-(\lambda+k+r)\tau_1} - e^{-(\lambda+k+r)\tau_2}\right] + \frac{kS_m}{r(\lambda + k)}\left[e^{-r\tau_1} - e^{-r\tau_2}\right]. \quad (3.15)$$

Again, when $S_m = 0$ with $k = -\alpha$ we get Eq. (3.14) above.

Box 3.4. Example: One barrel of oil extracted per year over 20 years. Assume that oil price follows an IGBM process. The initial price is 110 dollars per barrel, the (risk-neutral) long-term price $kS_m/(\lambda + k)$ is 90, $(\lambda + k) = 0.30$, and the riskless rate is 2 %. An annuity received continuously over this period is worth:

$$PV = \frac{110 - 90}{0.30 + 0.02}\left[1 - e^{-6.4}\right] + \frac{90}{0.02}\left[1 - e^{-0.4}\right] = 1,545.96\,\text{dollars}.$$

This amount can be compared to the initial investment required to extract one barrel of oil per year over the next 20 years.

In the particular case that $\tau_1 = 0$ we have:

$$PV = \frac{S_0 - \frac{kS_m}{\lambda+k}}{\lambda + k + r}\left[1 - e^{-(\lambda+k+r)\tau_2}\right] + \frac{kS_m}{r(\lambda + k)}\left[1 - e^{-r\tau_2}\right]. \tag{3.16}$$

If there is no risk then $\lambda = 0$; therefore:

$$PV = \frac{S_0 - S_m}{k + r}\left[1 - e^{-(k+r)\tau_2}\right] + \frac{S_m}{r}\left[1 - e^{-r\tau_2}\right].$$

On the other hand, if $\tau_2 \to \infty$ (with $\lambda \neq 0$) then:

$$PV = \frac{S_0 - \frac{kS_m}{\lambda+k}}{\lambda + k + r} + \frac{kS_m}{r(\lambda + k)}.$$

In this case, it can be observed that the project value is the sum of two components: one which is a function of the initial difference between the observed value and the long-term risk-neutral level of S, and another one related to the reversion value.

When the annuity is perpetual (i.e., it is a perpetuity) and S_t follows a GBM with $S_m = 0$ and $\alpha = -k$:

$$PV = \frac{S_0}{\lambda - \alpha + r}. \tag{3.17}$$

Besides, since $\alpha - \lambda = r - \delta$ then:

$$PV = \frac{S_0}{\delta}. \tag{3.18}$$

3.2.3 Futures Contracts Under the GBM

Now, the futures price $F(\cdot)$ (i.e. the value of the delivery price at time t such that the current value of the futures contract equals zero) is the expected spot price in a risk-neutral context. Besides, the properties of the log-normal distribution (S) imply that:

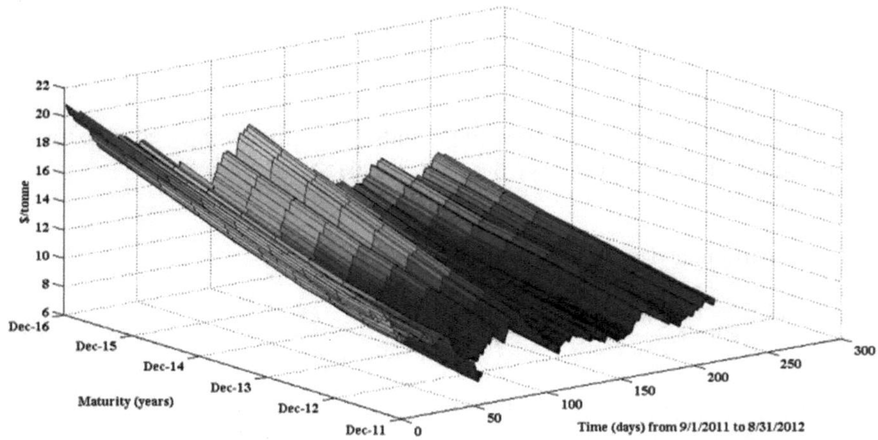

Fig. 3.3 Price surface for futures contracts on EU carbon emission allowances

$$F(S_0, t) = e^{E^Q(S)+\frac{1}{2}Var(S)} = e^{\ln S_0+\left(\alpha-\frac{\sigma^2}{2}-\lambda\right)t+\frac{\sigma^2}{2}t} = S_0 e^{(\alpha-\lambda)t}. \tag{3.19}$$

Stating this equation in logarithmic form we get:

$$\ln F(S_0, t) = \ln S_0 + (\alpha - \lambda)t.$$

Hence it could be possible to estimate the parameter $(\alpha - \lambda)$ in this equation from actual futures prices. S_0 could be taken directly from the spot market (assuming this exists). Figure 3.3 displays the surface of futures prices (on the ICE) regarding EU carbon emission allowances for different maturities; in this case, there is also a spot price.

3.2.4 Futures Contracts Under the IGBM

In order to value investments it can be convenient to use the prices on futures markets. These prices give the expected spot price in a risk-neutral world. For this purpose, the risk premium λS_t (which we assume to be proportional to S_t) is subtracted from the stochastic differential equation. This yields:

$$dS_t = [k(S_m - S_t) - \lambda S_t]dt + \sigma S_t dZ_t.$$

The expected value, or equivalently the futures price for maturity t, at time 0 is:

$$F(S_0, t) = E^Q(S_t) = S_0 e^{-(\lambda+k)t} + \frac{kS_m}{\lambda + k}\left[1 - e^{-(\lambda+k)t}\right]. \tag{3.20}$$

Thus $F(S_0, \infty) = kS_m/(\lambda + k)$ would be the long-term equilibrium price in the risk-neutral world and also on the futures market. There is a time $t^*_{1/2}$ at which the futures price reaches the mid value between the spot price S_0 and the equilibrium price in the long run $kS_m/(\lambda+k)$:

$$S_0 e^{-(\lambda+k)t^*_{1/2}} + \frac{kS_m}{\lambda+k}\left[1 - e^{-(\lambda+k)t^*_{1/2}}\right] = \frac{S_0 + \frac{kS_m}{\lambda+k}}{2}.$$

Hence we get:

$$\left(S_0 - \frac{kS_m}{\lambda+k}\right) e^{-(\lambda+k)t^*_{1/2}} = \frac{S_0 - \frac{kS_m}{\lambda+k}}{2}.$$

Similarly to the analysis of the spot price:

$$t^*_{1/2} = \frac{\ln 2}{k+\lambda}.$$

These formulas can be useful to check our numerical estimates of the underlying parameters. Assume there is a futures market with a high enough number of maturities available. Our estimate of $kS_m/(\lambda+k)$ can be easily checked since it should be the asymptotic value that distant futures prices tend to. On the other hand, if we find that two-year futures prices stand midway between the spot price and the equilibrium price, this means that $\lambda + k = (\ln 2)/2 = 0.3466$. In this regard, it is interesting to observe how futures prices change over time:

$$\frac{dF(S_0,t)}{dt} = (k+\lambda)\left[\frac{kS_m}{k+\lambda} - S_0\right] e^{-(k+\lambda)t}.$$

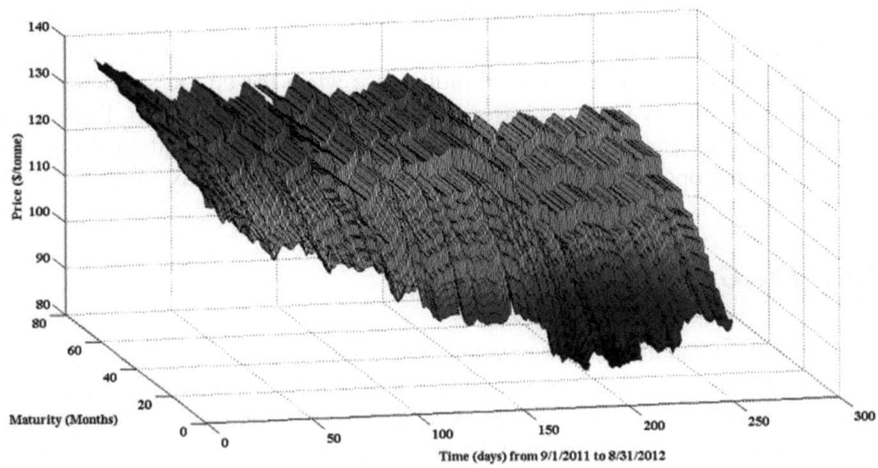

Fig. 3.4 Price surface for futures contracts on coal traded in Rotterdam

This instantaneous change depends on $k + \lambda$: if this sum is high, it takes less time to approach the equilibrium price. Figure 3.4 displays the surface of futures prices of ICE Rotterdam coal for different maturities.

In the case of the GBM with $S_m = 0$ and $k = -\alpha$ we have:

$$F(S_0, t) = S_0 e^{(\alpha-\lambda)t} = S_0 e^{(r-\delta)t}, \tag{3.21}$$

where δ denotes the convenience yield.

Estimation of the parameters in Eqs. (3.20) or (3.21) allows us to model the behavior of the futures market for terms or maturities that are well beyond those of available contracts (so we have no prices for the distant future). But there is a potential shortcoming: substituting the parameter estimates in Eqs. (3.20) or (3.21) can provide futures prices that are slightly different from actual prices of traded contracts. Nonetheless, this is not a major problem for valuing an annuity in the long run, since small differences tend to cancel each other. In the case of an IGBM process, from Eq. (3.20) we can deduce that $F(S_0, \infty) = kS_m/(\lambda + k)$ provided $k + \lambda > 0$. If k is very high, the limit equals S_m.

3.3 Fundamental Pricing Equation: The Perpetual Option

Here we deal with the valuation of the opportunity to invest in an asset whose price is governed by one of the above stochastic processes. This investment option is typically available only for a defined period. For example, our operations may be liable to a leasing contract with a known expiration date. Or the opportunity is up for grabs in the sense that if we do not exercise the option to invest (one or more of) our rivals will do thus displacing us.

Nonetheless, there can be instances in which the decision time frame is so long that it can be reasonably approached by the assumption of unlimited maturity. In this case it is possible to derive closed-form solutions for the value of the option. This case can be considered as the limiting case when the option's time to expiration tends to infinity (and used as such in numerical computations). In Chap. 4 we show the convergence of the numerical solutions toward the analytical solution of the perpetual option when the time to expiration is sufficiently long.

3.3.1 The GBM

Now, let $F(S, t)$ denote the market value of an asset that entitles the owner the full property of a project (say, a firm). S stands for the project's output price; assume it follows a GBM: $dS = \alpha S dt + \sigma S dZ$. Assume also that the project's output can itself be traded as an asset in financial markets. Of course, for investors to hold this asset, it has to provide a sufficiently high return. Let $\mu = \alpha + \delta$ be the total expected rate of return.

Following Dixit and Pindyck (1994), we find the value $F(S,t)$ by constructing a replicating portfolio, i.e., we combine traded assets of known value in such a way that the combination mimics the risk/return characteristics of $F(S,t)$. To be precise, we invest \$1 in the risk-free asset and purchase n units of the project's output. The cost of this portfolio is $(1+nS)$. Over a short time interval dt, the riskless asset pays rdt, while the output units provide a capital gain ndS and a dividend $n\delta Sdt$. The proportional rate of return on this portfolio is:

$$\frac{r + n\alpha S + n\delta S}{1+nS}dt + \frac{n\sigma S}{1+nS}dZ = \frac{r + n(\alpha+\delta)S}{1+nS}dt + \frac{n\sigma S}{1+nS}dZ.$$

As an alternative to the replicating portfolio, now consider the return from owning the project over the interval dt. Ownership has a cost $F(S,t)$. Barring any profit flow (for simplicity), the project only yields a capital gain dF, which can be derived through Ito's Lemma:

$$dF(S,t) = \left[\frac{\partial F}{\partial t} + \alpha S\frac{\partial F}{\partial S} + \frac{1}{2}\sigma^2 S^2 \frac{\partial^2 F}{\partial S^2}\right]dt + \sigma S\frac{\partial F}{\partial S}dZ.$$

In relative terms:

$$\frac{F_t + \alpha SF_S + \frac{1}{2}\sigma^2 S^2 F_{SS}}{F(S,t)}dt + \frac{\sigma SF_S}{F(S,t)}dZ.$$

By definition, the replicating portfolio has the same risk that owning the project. Therefore, the two coefficients of dZ in the two equations must be equal:

$$\frac{n\sigma S}{1+nS} = \frac{\sigma SF_S}{F(S,t)} \rightarrow \frac{n}{1+nS} = \frac{F_S}{F(S,t)} \rightarrow n^* = \frac{F_S}{F - SF_S}.$$

At the same time, the absence of arbitrage opportunities requires that two assets with identical risk earn equal return in the market:

$$\frac{r + n(\alpha+\delta)S}{1+nS} = \frac{F_t + \alpha SF_S + \frac{1}{2}\sigma^2 S^2 F_{SS}}{F(S,t)}.$$

Substituting n^* here, after some rearrangements and simplifications the return equation becomes a partial differential equation that the value must satisfy:

$$\frac{1}{2}\sigma^2 S^2 F_{SS} + (\alpha - \lambda)SF_S + F_t - rF = 0. \tag{3.22}$$

In our case of a perpetual option to invest in the project ($F_t = 0$), the option value $F(S,t)$ must satisfy the following differential equation, where sub-indices in S have been replaced by traditional derivative signs:

$$\frac{1}{2}\sigma^2 S^2 F'' + (\alpha - \lambda)SF' - rF = 0. \tag{3.23}$$

The solution to this equation is:

$$F(S) = A_1 S^{\gamma_1} + A_2 S^{\gamma_2}, \quad \gamma_1 > 0, \gamma_2 < 0. \tag{3.24}$$

Consider, for example, the option to invest in an industrial boiler. It may be argued that if the fuel price grows arbitrarily high the option will be worthless, so $A_1 = 0$ and hence:

$$F(S) = A_2 S^{\gamma_2}. \tag{3.25}$$

Taking the first and second derivatives, after substitution of F, F' and F'' in the differential equation we get a quadratic equation:

$$\frac{1}{2}\sigma^2\gamma_2^2 + \gamma_2\left(\alpha - \lambda - \frac{1}{2}\sigma^2\right) - r = 0. \tag{3.26}$$

The values of A_2 and the critical or trigger price S^* (below which to exercise of the option) remain to be determined. We resort to the value-matching condition and the smooth-pasting condition:

$$F(S^*) = V(S^*) - I(S^*). \tag{3.27}$$

$$F'(S^*) = V'(S^*) - I'(S^*). \tag{3.28}$$

3.3.2 Example 1: Optimal Timing Under Certainty (Finite-Lived Option)

Assume that the carbon allowance price (S) behaves deterministically over time:

$$dS_t = \alpha S_t dt \rightarrow S_t = S_0 e^{\alpha t},$$

and conforms exactly to the futures market curve ($\sigma = 0$ entails perfect foresight and no risk premium, $\lambda = 0$). We are free to choose the time to invest in a carbon capture unit over the first 10 years, i.e. the optimal time to invest $0 \leq T^* \leq 10$ must be determined (similarly to an American option). To this end, at any time we must assess whether it is better to invest immediately or rather to wait (for one more period).

Let I_T denote the investment outlay at time T that is required to save 1 tCO_2 per year from then onwards. We assume that this cost decreases at a constant rate β from current level I_0:

$$I_T = I_0 e^{-\beta T}, \quad \beta > 0.$$

We further assume that, upon investment at T, the savings (or sale revenues from spare allowances) will accrue over the next remaining years, $20 - T$; thus, cash inflows take place for a minimum of 10 years (if we invest at the end) up to a maximum of 20 years (if we invest now). Consider the following parameter values: $\alpha - \lambda = 0.05$, $r = 0.02$, $\beta = 0.025$, and $S_0 = 10$. The carbon price is thus expected to increase at a rate that is higher than the discount rate; this pushes for waiting to invest. Similarly, the investment cost decreases with time, which also calls for delaying investment. However, postponing the investment implies a lower number of years for profiting from the investment, so there is clearly a trade-off. The (time-T) net present value of investing at T is given by:

$$V(S_T) - I_T = \frac{S_T}{\alpha - \lambda - r}\left[e^{(\alpha-\lambda-r)(20-T)} - 1\right] - I_0 e^{-\beta T}.$$

Seen from today, the time-0 net present value amounts to:

$$NPV = [V(S_T) - I_T]e^{-rT} = \frac{S_0 e^{(\alpha-\lambda-r)T}}{\alpha - \lambda - r}\left[e^{(\alpha-\lambda-r)(20-T)} - 1\right] - I_0 e^{-(\beta+r)T}. \quad (3.29)$$

With the above figures, under immediate investment ($T = 0$) we get a gross value:

$$V(S_0) = \frac{10}{0.05 - 0.02}\left[e^{(0.05-0.02)20} - 1\right] = 274.04.$$

Note that if $\alpha - \lambda - r > 0$ then $S_0 e^{(\alpha-\lambda-r)T}$ increases with T, while both $e^{(\alpha-\lambda-r)(20-T)}$ and $I_0 e^{-(\beta+r)T}$ decrease with T.

In general, the optimal time to invest T^* under certainty can be determined by differentiating Eq. (3.29) above with respect to T and setting it equal to zero:

$$S_0 e^{(\alpha-\lambda-r)T^*} = (\beta + r)I_0 e^{-(\beta+r)T^*} \rightarrow e^{(\alpha-\lambda+\beta)T^*} = \frac{(\beta + r)I_0}{S_0} \rightarrow$$

$$T^* = \frac{\ln[(\beta + r)I_0] - \ln S_0}{\alpha - \lambda + \beta}. \quad (3.30)$$

For $I_0 = 260$ we compute $T^* = 2.09$. At that precise time we have:

$$S_{T^*} = S_0 e^{(\alpha-\lambda)T^*} = 11.10 \rightarrow V(S_{T^*}) = \frac{S_{T^*}}{\alpha - \lambda - r}\left[e^{(\alpha-\lambda-r)(20-T^*)} - 1\right] = 263.22;$$

$$I_{T^*} = I_0 e^{-\beta T^*} = 246.74.$$

Hence we derive: $NPV = (263.22 - 246.74)e^{-0.02\times 2.09} = 15.80$, which is the maximum value.

At this point an important remark is in order: the NPV of investing immediately amounts to $V(S_0) - I_0 = 274.04 - 260 = 14.04 > 0$. Yet we have just concluded that, even under certainty, waiting (until T^*) is the best course of action: the NPV reaches its highest level 15.80. It can be easily checked that this is indeed the

Table 3.1 Investment timing and the NPV

Time T (years)	NPV
0	14.0396
1	15.3287
2	15.8053
2.0934	15.8088
3	15.4820
5	12.4806
10	−8.3633

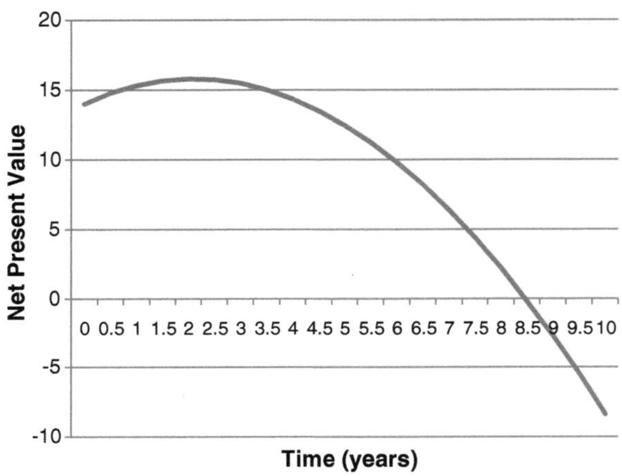

Fig. 3.5 The NPV as a function of the time of investment

optimum; see Table 3.1. As a general rule, it is convenient to test that there are no corner solutions when $T = 0$ or $T = 10$. Figure 3.5 shows the NPV for different times to invest T.

3.3.3 Example 2: Optimal Time to Invest Under a GBM

Constant investment cost It is hard to think that we have the option to undertake an investment at any time into the most distant future. Nonetheless, if the option's time to expiration is long enough, the perpetual option, which has an analytic solution, can be a good approximation and a benchmark for testing the reliability of valuations under finite lives.

Let H denote the value of a perpetual option to invest in an asset worth V. The underlying asset price follows:

$$dS_t = (\alpha - \lambda)S_t dt + \sigma S_t dZ_t.$$

We know that the option value must satisfy the following differential equation:

$$\frac{1}{2}\sigma^2 S^2 H'' + (\alpha - \lambda)SH' - rH = 0.$$

The solution to this equation is:

$$H(S) = A_1 S^{\gamma_1} + A_2 S^{\gamma_2}, \quad \gamma_1 > 0, \gamma_2 < 0.$$

Consider, for example, the case that the investment, once undertaken, saves 1 tCO_2 per year over the next 20 years. Thus, the firm has a flow of revenues over 20 years. Since the price of a carbon allowance is stochastic, the investment option's value depends on stochastic revenues. Then, if S falls to zero the option must be worthless (since S will remain at that level forever); thus: $H(0) = 0$, which implies $A_2 = 0$. Hence:

$$H(S) = A_1 S^{\gamma_1} \tag{3.31}$$

Consequently:

$$H' = \gamma_1 A_1 S^{\gamma_1 - 1}, \quad H'' = \gamma_1(\gamma_1 - 1)A_1 S^{\gamma_1 - 2}.$$

The differential equation reduces to:

$$\frac{1}{2}\sigma^2 \gamma_1(\gamma_1 - 1) + (\alpha - \lambda)\gamma_1 - r = 0. \tag{3.32}$$

With $\alpha - \lambda = 0.05$, $r = 0.10$, and $\sigma = 0.40$, we get $\gamma_1 = 1.3211 > 0$.

On the other hand, we have two conditions for determining the values of A_1 and S^*. The value-matching condition states that:

$$H(S^*) = V(S^*) - I(S^*) \rightarrow A_1(S^*)^{\gamma_1} = \frac{S^*}{\alpha - \lambda - r}\left[e^{(\alpha - \lambda - r)20} - 1\right] - I.$$

With the above figures we get:

$$A_1(S^*)^{\gamma_1} = 12.6424 S^* - I. \tag{3.33}$$

As for the smooth-pasting condition, it establishes that:

$$\gamma_1 A_1 (S^*)^{\gamma_1 - 1} = 12.6424. \tag{3.34}$$

Since γ_1 is already known, we have a set of two linearly independent equations that allow us to determine both A_1 and $S^*$for any given value of I. Solving them for A_1 and then equating yields:

$$12.6424 S^* \left(1 - \frac{1}{\gamma_1}\right) = I \rightarrow I = 3.0731 S^*.$$

This means that there is a boundary or locus in the (I, S) space that separates the region where immediate investment is optimal (the investment region) from that in which it is better to wait (the continuation region); see Fig. 3.6. In particular,

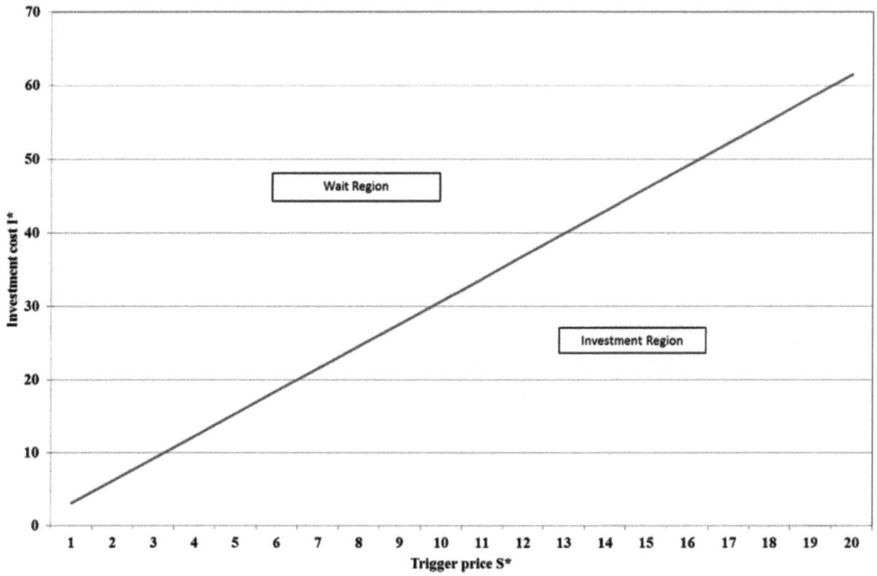

Fig. 3.6 Optimal boundary to invest in a project that saves 1 tCO_2 per year over 20 years

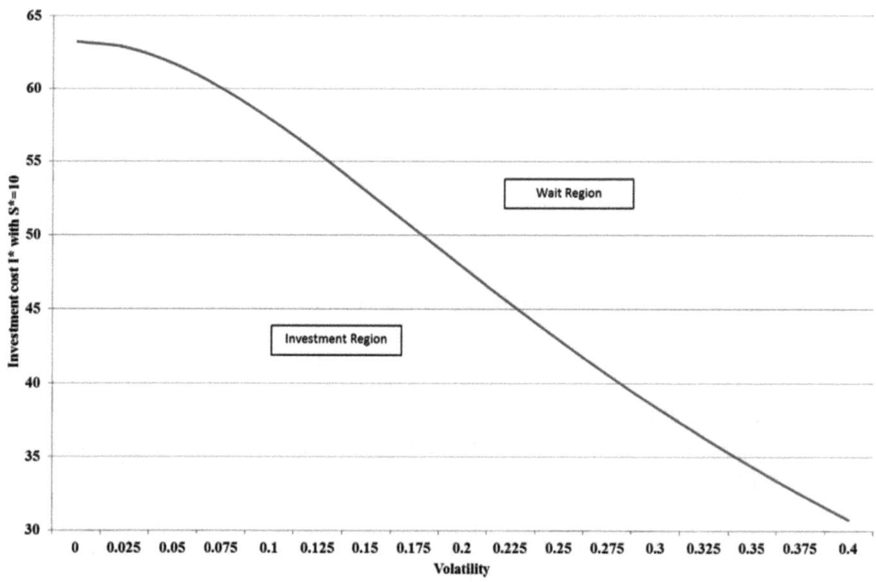

Fig. 3.7 Optimal boundary to invest when S* = 10 as a function of carbon price volatility

Table 3.2 Sensitivity of the option value to volatility

	$\sigma = 0$	$\sigma = 0.10$	$\sigma = 0.20$	$\sigma = 0.30$	$\sigma = 0.40$
γ_1	2.00	1.84	1.60	1.43	1.32
A_1	0.63	0.98	1.93	3.22	4.56
$A_1 S^{\gamma_1}$	63.21	68.55	78.60	88.03	95.69
I	63.21	57.88	47.83	38.40	30.73

investment should take place as soon as $3.0731S^* \geq I$. For $S^* = 10$ the investment threshold is 30.7314; from the smooth-pasting condition the unknown coefficient is:

$$A_1 = \frac{12.6424}{\gamma_1 (S^*)^{\gamma_1 - 1}} = 4.5681,$$

and we can check that the value-matching condition is satisfied. See Fig. 3.7.

It is interesting to analyze the impact of volatility on the solution; assume $S_0 = S^* = 10$. See Table 3.2. The second column ($\sigma = 0$) represents the case under certainty. A higher volatility enhances the option value: it rises from 63.21 (under $\sigma = 0$) to 95.6927 (with $\sigma = 0.40$). Since we assume a given value of S^*, as volatility increases the investment threshold drops: it falls from 63.21 (under $\sigma = 0$) to 30.73 (with $\sigma = 0.40$); thus investing becomes more demanding.

It is also possible to conduct the opposite analysis, i.e., assumed a known investment outlay I, determine the critical or trigger price S^* for investing immediately. In doing so, our computation of the coefficient A_1 would depend on I. Assume, for example, that $I = 280$. In this case we get $S^* = 91.1119$ and $A_1 = 2.2468$ while $\gamma_1 = 1.3211$ remains the same. It is easy to check that the two boundary conditions hold:

Value-matching: $A_1 (S^*)^{\gamma_1} = 12.6424S^* - I \quad (= 871.874)$.
Smooth-pasting: $\gamma_1 A_1 (S^*)^{\gamma_1 - 1} = 12.6424$.
And that both values fall along the frontier between the continuation region and the investment region: $I = 3.0731\, S^* = 3.0731 \times 91.1119 = 280$.

When the profits from the investment are perceived indefinitely (instead of over 20 years), the (gross) present value of the investment is given by:

$$V(S_T) = \frac{S_0}{r - (\alpha - \lambda)} = \frac{S_0}{\delta}.$$

This is the case described in Dixit and Pindyck (1994, pp. 182–184); the above example is therefore a more general case, with an infinite-lived investment option but investment profits over a finite period.

Variable (deterministic) investment cost In this case we have:

$$dI_t = \beta I_t dt \rightarrow I_t = I_0 e^{\beta t}.$$

The perpetual option must satisfy the differential equation:

$$\frac{1}{2}\sigma^2 S^2 H_{SS} + (\alpha - \lambda)SH_S + \beta I H_I - rH = 0, \tag{3.35}$$

along with the following boundary conditions:

$$H(S^*, I) = V(S^*) - I(S^*) = 12.6424S^* - I, \tag{3.36}$$

$$H_S(S^*, I) = 12.6424, \tag{3.37}$$

$$H_I(S^*, I) = -1. \tag{3.38}$$

Following Dixit and Pindyck (1994, p. 210), doubling S and I will double the value of the project and also the cost of investing. Correspondingly, the value of the option should be homogenous of degree 1 in (S, I), enabling us to write:

$$H(kS, kI) = kH(S, I) \rightarrow H(S, I) = \frac{1}{k}H(kS, kI).$$

Hence, setting $k = 1/I$ we can write:

$$H(S, I) = IH\left(\frac{S}{I}, \frac{I}{I}\right) = Ih\left(\frac{S}{I}\right). \tag{3.39}$$

The partial derivatives in the differential equation then become:

$$H_S = h'\left(\frac{S}{I}\right), \quad H_{SS} = \frac{1}{I}h''\left(\frac{S}{I}\right), \quad H_I = h\left(\frac{S}{I}\right) - \frac{S}{I}h'\left(\frac{S}{I}\right).$$

Now substituting above, dividing throughout by I, and defining $x \equiv S/I$ we get:

$$\frac{1}{2}\sigma^2 x^2 h''(x) + (\alpha - \lambda - \beta)xh'(x) + (\beta - r)h(x) = 0, \tag{3.40}$$

where x is now the function to be determined.

Let the solution be:

$$h(x) = A_1 x^{\gamma_1} + A_2 x^{\gamma_2}, \quad \gamma_1 > 0,\ \gamma_2 < 0. \tag{3.41}$$

In the case that revenues (S) are stochastic, if $x \rightarrow 0$ the option to invest in carbon capture should be worthless, so $A_2 = 0$. Thus:

$$h(x) = A_1 x^{\gamma_1}, \quad \gamma_1 > 0.$$

The boundary conditions will be:

Value-matching: $h(x^*)^{\gamma_1} = 12.6424x^* - 1$.
Smooth-pasting: $h'(x^*)^{\gamma_1} = 12.6424$.
Given that we know $h(x) = A_1 x^{\gamma_1}$ we derive:
From value-matching: $A_1 = (12.6424x^* - 1)(x^*)^{-\gamma_1}$.
From smooth-pasting: $A_1 = \frac{12.6424}{\gamma_1}(x^*)^{-\gamma_1+1}$.

Table 3.3 Sensitivity of the option value to volatility ($I_0 = 280$)

	$\sigma = 0$	$\sigma = 0.10$	$\sigma = 0.20$	$\sigma = 0.30$	$\sigma = 0.40$
γ_1	1.6667	1.6023	1.4781	1.3663	1.2816
x^*	0.1977	0.2104	0.2445	0.2950	0.3599
S^*	55.3692	58.9180	68.4645	82.6043	100.785

Equating both expressions for A_1:

$$x^* \equiv \left(\frac{S}{I}\right)^* = \frac{\gamma_1}{12.6424(\gamma_1 - 1)}.$$

On the other hand, substituting $h(x)$ in the differential equation yields:

$$\frac{1}{2}\sigma^2\gamma_1^2 + \left(\alpha - \lambda - \beta - \frac{1}{2}\sigma^2\right)\gamma_1 + (\beta - r) = 0,$$

which allows to compute the power γ_1.

Consider the following parameter values: $\alpha - \lambda = 0.05$, $r = 0.10$, $\sigma = 0.40$, and $\beta = -0.025$ (i.e., decreasing investment cost). We get: $\gamma_1 = 1.2816$, and hence $x^* = 0.3599$. Therefore, for an initial outlay $I_0 = 280$ the trigger price is $S^* = 100.7858$. The value of the option at the initial time turns out to be:

$$H = 12.6424 \times 100.7858 - 280 = 994.175.$$

This coincides with the NPV at time 0 (at S^*, it is optimal to invest).

We can further check the impact of volatility. See Table 3.3. The case under certainty corresponds to $\sigma = 0$. As we move from the deterministic case to greater levels of uncertainty the critical prices rises, i.e., the conditions for optimally exercising the option to invest get more stringent.

For example, in the case $\sigma = 0$, the differential equation simplifies to:

$$(\alpha - \lambda - \beta)xh'(x) + (\beta - r)h(x) = 0.$$

Substituting $h(x)$ and $h'(x)$ yields:

$$\gamma_1 = \frac{r - \beta}{(\alpha - \lambda - \beta)} = 1.6667.$$

Hence $A_1 = 22.3481$ and $x^* = 0.1977$. Consequently, $S^* = 280\ x^* = 55.3692$.

3.3.4 Example 3: Two correlated GBMs

Now we consider two correlated GBMs that stand for the underlying asset price and the investment cost (in the risk-neutral world):

$$dS_t = (\alpha_S - \lambda_S)S_t dt + \sigma_S S_t dW_t^S, \tag{3.42}$$

$$dI_t = (\alpha_I - \lambda_I)S_t dt + \sigma_I I_t dW_t^I, \tag{3.43}$$

$$E\left(dW_t^S, dW_t^I\right) = \rho dt. \tag{3.44}$$

Upon investment at T, the net value of the project is $V(S_T) - I_T$, or:

$$\frac{S_T}{\alpha_S - \lambda_S - r}\left[e^{(\alpha_S - \lambda_S - r)20} - 1\right] - I_T.$$

The boundary conditions are:

Value-matching: $H(S^*, I^*) = V(S^*) - I^* = \frac{S_T}{\alpha_S - \lambda_S - r}\left[e^{(\alpha_S - \lambda_S - r)20} - 1\right] - I_T$.
Smooth-pasting: $H_S(S^*, I^*) = \frac{e^{(\alpha_S - \lambda_S - r)20} - 1}{\alpha_S - \lambda_S - r}$, and $H_I(S^*, I^*) = -1$.
Following Dixit and Pindyck (1994), define $x \equiv S/I$. Then $H(S, I) = Ih(x)$, and:
Value-matching: $h(x) = \frac{x}{\alpha_S - \lambda_S - r}\left[e^{(\alpha_S - \lambda_S - r)20} - 1\right] - 1 = 12.6424x - 1$.
Smooth-pasting: $h'^{(x)} = \frac{e^{(\alpha_S - \lambda_S - r)20} - 1}{\alpha_S - \lambda_S - r} = 12.6424$, and $h(x) - xh'(x) = -1$.

The value of γ must satisfy the quadratic equation:

$$\frac{1}{2}\left(\sigma_S^2 - 2\rho\sigma_S\sigma_I + \sigma_I^2\right)\gamma(\gamma - 1) + [(\alpha_S - \lambda_S) - (\alpha_I - \lambda_I)]\gamma + (\alpha_I - \lambda_I) - r = 0,$$

or

$$\frac{1}{2}\left(\sigma_S^2 - 2\rho\sigma_S\sigma_I + \sigma_I^2\right)\gamma^2 + \left[(\alpha_S - \lambda_S) - (\alpha_I - \lambda_I) - \frac{1}{2}\left(\sigma_S^2 + \sigma_I^2\right) + \rho\sigma_S\sigma_I\right]\gamma + (\alpha_I - \lambda_I) - r = 0,$$

where γ_1 denotes the positive root.

Since $h(x) = A_1 x^{\gamma_1}$ and $h'(x) = \gamma_1 A_1 x^{\gamma_1 - 1}$, it can be checked that the straight line separating the continuation region from the investment region is given by:

$$x^* \equiv \frac{S^*}{I^*} = \frac{\gamma_1}{12.6424(\gamma_1 - 1)}.$$

See Fig. 3.8.

Assume that $\alpha_S - \lambda_S = 0.05$, $\sigma_S = 0.40$, $\alpha_I - \lambda_I = 0.03$, $\sigma_I = 0.20$, $r = 0.10$, and $\rho = 0.50$. Under these circumstances we get: $\gamma_1 = 1.4637$, and hence $x^* = 0.2496$ and $A_1 = 16.4370$. They imply $h(x^*) = A_1(x^*)^{\gamma_1} = 2.1546$. This amount in turn equals $12.6424x^* - 1 = 2.1546$, so we check that the value-matching condition holds. In addition, $h'(x^*) = \gamma_1 A_1 (x^*)^{\gamma_1 - 1} = 1.4637 \times 16.4370 \times (0.2496)^{0.4637} = 12.6424$ so the smooth-pasting condition also applies. The value of the option, assuming $I = 280$ and $S = 10$, would be $H(S, I) = Ih(x) = IA_1(10/280)^{\gamma_1} = 35.05$.

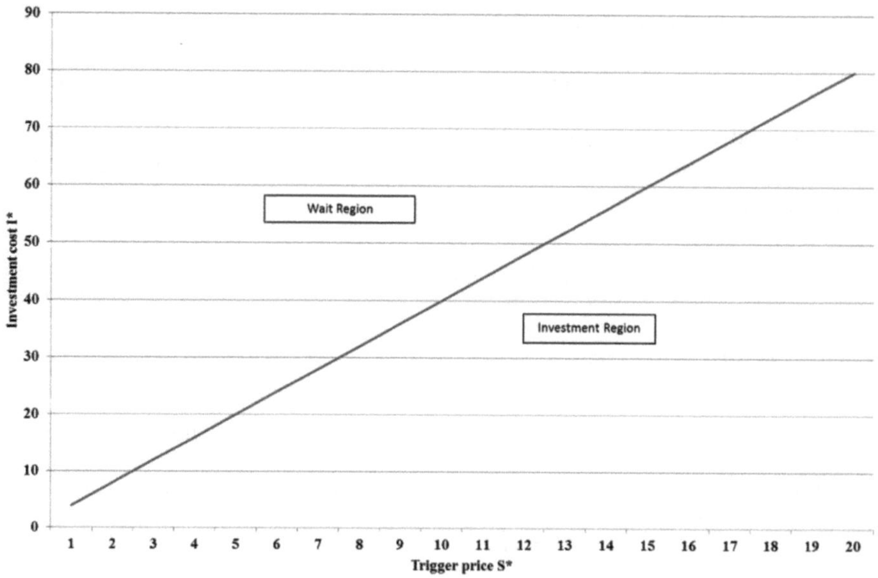

Fig. 3.8 Optimal boundary under two correlated GBMs

3.3.5 The IGBM

Next we want to derive the value H of an opportunity to invest in a project whose value V in turn depends on an asset whose price S follows an IGBM process. In general H will depend on S and t. If certain "complete market" assumptions hold, then it can be shown that its value will satisfy the differential equation:

$$\frac{1}{2}\sigma^2 S^2 \frac{\partial^2 H}{\partial S^2} + [k(S_m - S) - \lambda S]\frac{\partial H}{\partial S} + \frac{\partial H}{\partial t} - rH = 0. \tag{3.45}$$

If the investment option's time to maturity is infinite then the term $\partial H/\partial t = 0$ disappears in Eq. (3.45), which now can be expressed as:

$$\frac{1}{2}\sigma^2 S^2 H'' + [k(S_m - S) - \lambda S]H' - rH = 0. \tag{3.46}$$

This equation may be rewritten as:

$$S^2 H'' + (\alpha S + \beta)H' - \gamma H = 0, \tag{3.47}$$

where the following notation has been adopted:

$$\alpha \equiv \frac{2(k+\lambda)}{\sigma^2}, \quad \beta \equiv \frac{2kS_m}{\sigma^2}, \quad \gamma \equiv \frac{2r}{\sigma^2}.$$

In order to find a solution to this equation, we define a function $h(\beta S^{-1})$ by

$$H(S) = A_0\left(\beta S^{-1}\right)^{\theta} h\left(\beta S^{-1}\right), \tag{3.48}$$

where A_0 and θ are constants that will be chosen so as to make $h(\cdot)$ satisfy a differential equation with a known solution. The first and second derivatives, divided by $A_0\beta^{\theta}$, are:

$$\frac{H'(S)}{A_0\beta^{\theta}} = (-\theta)S^{-\theta-1}h\left(\beta S^{-1}\right) + S^{-\theta-2}h'\left(\beta S^{-1}\right)(-\beta),$$

$$\begin{aligned}\frac{H''(S)}{A_0\beta^{\theta}} &= \theta(\theta+1)S^{-\theta-2}h\left(\beta S^{-1}\right) + (-\theta)S^{-\theta-3}h'\left(\beta S^{-1}\right)(-\beta)\\ &\quad + (-\theta-2)S^{-\theta-3}h'\left(\beta S^{-1}\right)(-\beta) + S^{-\theta-4}h''\left(\beta S^{-1}\right)\beta^2.\end{aligned}$$

Substituting these expressions in Eq. (3.47) and simplifying we get:

$$\begin{aligned}&(\theta(\theta+1) - \alpha\theta - \gamma)S^{-\theta}h\left(\beta S^{-1}\right) + S^{-\theta-1}[S^{-1}\beta^2 h''\left(\beta S^{-1}\right)\\ &+h'\left(\beta S^{-1}\right)\left(\beta\theta + (\theta+2)\beta - \alpha\beta - S^{-1}\beta^2\right) - h\left(\beta S^{-1}\right)\beta\theta] = 0.\end{aligned}$$

For this equality to hold, first it must be:

$$\theta(\theta+1) - \alpha\theta - \gamma = \theta^2 + \theta(1-\alpha) - \gamma = 0. \tag{3.49}$$

This equation allows to determine the positive value of θ, since the remaining terms are known constants. Once the value of θ has been obtained, the remainder of the equation is:

$$\left(\beta S^{-1}\right)h''\left(\beta S^{-1}\right) + h'\left(\beta S^{-1}\right)\left(2\theta + 2 - \alpha - \beta S^{-1}\right) - \theta h\left(\beta S^{-1}\right) = 0. \tag{3.50}$$

This is Kummer's Differential Equation, where: $a = \theta$, $b = 2\theta + 2 - \alpha$, and $z = \left(\beta S^{-1}\right)$. The general solution to this equation has the form:

$$h\left(\beta S^{-1}\right) = A_1 U(a,b,z) + A_2 M(a,b,z). \tag{3.51}$$

where $U(a,b,z)$ is Tricomi's or second-order hypergeometric function, and $M(a,b,z)$ is Kummer's or first-order hypergeometric function.

Therefore, the general solution to $H(S)$ will be:

$$H(S) = A_0\left(\beta S^{-1}\right)^{\theta}[A_1 U(a,b,z) + A_2 M(a,b,z)]. \tag{3.52}$$

The second-order hypergeometric function has the following representation:

$$U(a,b,z) = \frac{\Gamma(1-b)}{\Gamma(1+a-b)}M(a,b,z) + \frac{\Gamma(b-1)}{\Gamma(a)z^{b-1}}M(1+a-b, 2-b, z),$$

where $\Gamma(\cdot)$ is the gamma function and the value of M(a, b, z) is given by:

$$M(a,b,z) = 1 + \frac{a}{b}z + \frac{a(a+1)}{b(b+1)}\frac{z^2}{2!} + \frac{a(a+1)(a+2)}{b(b+1)(b+2)}\frac{z^3}{3!} + \cdots$$

The derivatives of Kummer's function have the following properties:

$$\frac{\partial M(a,b,z)}{\partial z} = \frac{a}{b}M(a+1,b+1,z),$$

$$\frac{\partial^2 M(a,b,z)}{\partial z^2} = \frac{a(a+1)}{b(b+1)}M(a+2,b+2,z).$$

The derivatives of Tricomi's function satisfy:

$$\frac{\partial U(a,b,z)}{\partial z} = -aU(a+1,b+1,z),$$

$$\frac{\partial^2 U(a,b,z)}{\partial z^2} = a(a+1)U(a+2,b+2,z).$$

The boundary conditions will determine whether A_1 or A_2 in Eq. (3.19) are zero. If, for example, S refers to a fuel input the firm faces stochastic costs. An upward evolution in S entails a reduction in profits, so $F(\infty) = 0$ and $z = 0$, then $A_1 = 0$ and the term in Kummer's function remains. The solution is:

$$H(S) = A_m\left(\beta S^{-1}\right)^{\theta} M(a,b,z), \tag{3.53}$$

with $A_m \equiv A_0 A_2$. The constant A_m and the critical value S^* below which it is optimal to invest must be jointly determined by the remaining two boundary conditions:

(a) Value-Matching:

$$H(S^*) = V(S^*) - I(S^*), \tag{3.54}$$

(b) Smooth-Pasting:

$$H'(S^*) = V'(S^*) - I'(S^*). \tag{3.55}$$

3.3.6 *Example 4: Optimal Time to Invest Under an IGBM*

Consider a potential investment at time T that consumes 1 barrel of oil per day from time τ_1 to τ_2. The present value of fuel costs in this project amounts to:

$$PV = \frac{S_T - \frac{kS_m}{\lambda+k}}{\lambda+k+r}\left[e^{-(\lambda+k+r)\tau_1} - e^{-(\lambda+k+r)\tau_2}\right] + \frac{kS_m}{r(\lambda+k)}\left[e^{-r\tau_1} - e^{-r\tau_2}\right]. \tag{3.56}$$

This value can be expressed as $x_0 + x_1 S_T$, i.e., as a linear function of S_T, with:

$$x_0 \equiv -\frac{\frac{kS_m}{\lambda+k}}{\lambda+k+r}\left[e^{-(\lambda+k+r)\tau_1} - e^{-(\lambda+k+r)\tau_2}\right] + \frac{kS_m}{r(\lambda+k)}\left[e^{-r\tau_1} - e^{-r\tau_2}\right],$$

$$x_1 \equiv \frac{e^{-(\lambda+k+r)\tau_1} - e^{-(\lambda+k+r)\tau_2}}{\lambda+k+r}.$$

Now assume the following parameter values: $\frac{kS_m}{\lambda+k} = 90$, $\lambda + k = 0.30$, $\tau_1 = 0$, $\tau_2 = 20$, $r = 0.02$, and $\sigma = 0.25$. Under these circumstances the present value of fuel costs is given by:

$$PV = 1,202.7771 + 3.1198 S_T.$$

Thus, for $S_T = 110$ in particular, we would get $PV = 1,545.96$. If net revenues were 1,500.00 then the investment would have a net present value $NPV = 297.2229 - 3.1198 S_T$.

The boundary conditions will be:
Value-matching:

$$H(S^*) = A_m\left(\beta(S^*)^{-1}\right)^{\theta} M\left(a, b, \beta(S^*)^{-1}\right) = 297.2229 - 3.1198 S^*.$$

Smooth-pasting:

$$H'(S^*) = A_m\left(\beta(S^*)^{-1}\right)^{\theta}\left(-\theta(S^*)^{-1} M\left(a, b, \beta(S^*)^{-1}\right) - \frac{a\beta}{b(S^*)^2} M\left(a+1, b+1, \beta(S^*)^{-1}\right)\right) = -3.1198.$$

By computing $\alpha = -9.6$, $\beta = 864$, $\gamma = 0.64$, and $\theta = 0.060037$, we derive $a = 0.060037$ and $b = 11.72$. We thus have a system of two equations that will allow us determine A_m and S^*. From value-matching we get:

$$A_m\left(\beta(S^*)^{-1}\right)^{\theta} = \frac{297.2229 - 3.1198 S^*}{M\left(a, b, \beta(S^*)^{-1}\right)}.$$

We substitute this in the smooth-pasting condition, which results in a single equation in one unknown, S^*; the solution is $S^* = 52.482$. Next it is easy to derive $A_m = 81.9573$. Finally, the value of the option is given by:

$$H(S) = 81.9573\left(\frac{864}{S}\right)^{0.060037} M\left(0.060037,\ 11.72,\ \frac{864}{S}\right).$$

We can check the fulfillment of the value matching condition $H(S^*) = V(S^*) - I(S^*)$:

$$H(52.482) = 133.4891 = 297.2229 - 3.1198 \times 52.482.$$

Similarly, the smooth-pasting condition $H'(S^*) = V'(S^*) - I'(S^*)$ is satisfied:

$$H'(52.482) = -3.1198.$$

In our case, for $S_0 = 110$ the option value is $H = 98.5127$. Instead, $NPV = 297.2229 - 3.1198 \times 110 = -45.9559$. Thus, it is optimal to wait since the current costs are high and they will presumably fall because of mean reversion.

If there were no other option but to invest now or never, the hurdle point would be $V(S^{**}) = I(S^{**})$, when we get $S^{**} = 297.2229/3.1198 = 95.2696$. However, with a perpetual option it is preferable to wait, since in principle and in the long run the price is going to decrease and fluctuate around a level $kS_m/\lambda + k = 90$, and then keep on waiting until it reaches $S^* = 52.482$ or below this value, so that the option value equals the net value of the investment. When the option may be exercised only during a finite period, the threshold S^* will take on a value between 52.482 and 95.2699.

3.4 Pricing Formulas for European Options

Unlike American options in general, which lack an explicit valuation formula, it is possible to derive exact pricing formulas for European options. Assume that the price of the underlying asset follows a GBM. Black and Scholes (1973) and Merton (1973) showed that the value of a call option (C) and a put option (P) with exercise price K and maturity T (in years) are given by:

$$C = S_0 e^{((\alpha-\lambda)-r)T} N(d_1) - Ke^{-rT} N(d_2), \tag{3.57}$$

$$P = Ke^{-rT} N(-d_2) - S_0 e^{((\alpha-\lambda)-r)T} N(-d_1), \tag{3.58}$$

where

$$d_1 = \frac{\ln\frac{S_0}{K} + \left[(\alpha-\lambda) + \frac{1}{2}\sigma^2\right]T}{\sigma\sqrt{T}}, \quad d_2 = 1 - d_1 = \frac{\ln\frac{S_0}{K} + \left[(\alpha-\lambda) - \frac{1}{2}\sigma^2\right]T}{\sigma\sqrt{T}}.$$

References

Baker MP, Mayfield ES, Parsons JE (1998) Alternative models of uncertain commodity prices for use with modern asset pricing methods. Energy J 19(1):115–148

Benth FE, Kiesel R, Nazarova A (2012) A critical empirical study of three electricity spot price models. Energy Econ 34:1589–1616

Black F, Scholes MB (1973) The pricing of options and corporate liabilities. J Polit Econ 81(3):637–654

Cortazar G, Schwartz ES (2003) Implementing a stochastic model for oil futures prices. Energy Econ 25:215–238
Dixit AK, Pindyck RS (1994) Investment under uncertainty. Princeton University Press, Princeton, NJ
Keles D, Genoese M, Möst D, Fichtner W (2012) Comparison of extended mean-reversion and time series models for electricity spot price simulation considering negative prices. Energy Econ 34:1012–1032
Kloeden PE, Platen E (1992) Numerical solution of stochastic differential equations. Springer
Merton RC (1973) Theory of rational option pricing. Bell J Econ Manage Sci 4(1):141–183
Nomikos N, Andriosopoulos K (2012) Modelling energy spot prices: empirical evidence from NYMEX. Energy Econ 34:1153–1169
Pilipovic D (1998) Energy risk. McGraw-Hill
Ronn EI (2002) Real options and energy management. Risk Books
Schwartz ES (1997) The stochastic behavior of commodity prices: implications for valuation and hedging. J Finance 52(3):923–973

Chapter 4
Binomial Lattices

4.1 Introduction

The value of an asset stems from its future cash flows. As long as the latter are certain, the valuation problem comes down to finding the appropriate discount factors to translate these future flows into their present equivalents. Uncertainty in future cash flows, however, rises two issues. First, the present value (*PV*) of future cash flows depends not only on the time value of money but on investors' appetite for risk as well. Thus, either discount factors are enhanced to take account also of suitable risk premiums, or we stick to the former discount factors but replacing the anticipated flows by their certainty equivalents. Second, deriving risk premiums becomes burdensome when the cash flows are a non-linear function of the underlying stochastic variable. In such a case, computing certainty equivalents is the only viable route in practice.

This is not much of a problem after all when the non-linear cash flows depend exclusively on past information. Consider, for example, a European call option; it is the right to buy a given amount of an asset at a given price (the exercise price or the striking price, *K*) on a specific date (at maturity, *T*). Its value at that time is the maximum of two amounts: $max(S_T - K, 0)$. The difference $S_T - K$ is clearly dependent on the past (the asset price reaches a particular terminal node because there has been a particular number of upward and downward jumps previously; otherwise the price would end up at other node). This is just an example of forward induction. Monte Carlo simulation is well suited for valuing assets whose cash flows display this characteristic.

Matters get more complicated when the non-linear cash flows at a given time depend on future information. American options, which can be exercised at any time up to expiration, are one such case. Thus, consider a dual-fuel boiler which can run on either coal or biomass at any time but incurs switching costs. In principle it is not obvious whether we should start today burning biomass or coal; and, of course, this affects the boiler's *PV*. Rational decision making under these circumstances calls for taking into account not only past information but also expectations on future events. In this case one sensible approach is to map all the

L.M. Abadie and J.M. Chamorro, *Investment in Energy Assets Under Uncertainty*, Lecture Notes in Energy 21, DOI: 10.1007/978-1-4471-5592-8_4,

possibilities that can arise as the future unfolds, take the best decision in each case, and then adopt some type of backward induction. Starting from the option's maturity and proceeding backwards allows us to draw the sequence of optimal decisions under any contingency; eventually the present date is reached, which helps to elucidate if it is better to start running on coal or biomass. Needless to say, deciding optimally is a pre-requisite for maximizing the value of the option. In other words, the option holder maximizes its value as long as the boiler starts in the optimal mode.

There are several backward induction methods, among them binomial lattice; we now turn to them. Their basic premise is that uncertainty at any time can be represented through two alternative states (of nature). A binary distribution (or Bernoulli distribution) is a discrete distribution which can take on two values, 1 and 0 (or white ball and black ball), with probabilities q and $(1-q)$, respectively. The mathematical expectation is q, while the variance is $q(1-q)$.

Let $X_1, X_2, \ldots, X_n$ be independent random variables following a binary distribution with probability q. The random variable Z of the binomial distribution with parameters q and n (for short, $b(q,n)$) is defined by:

$$Z = X_1 + X_2 + \ldots + X_n.$$

Thus Z can be interpreted as the total outcome after n drawings or experiments. Since each term in the sum can only take on the values 0 and 1, the values for the binomial variable $b(q, n)$ that are possible are $\{0, 1, 2, \ldots, n\}$. Given that we do not know in advance what value Z will have for each experiment $n > 0$, to us Z is a stochastic process. See Shreve (2005).

4.2 The Basic Setting: Binomial Lattice Under a GBM

Let S denote the current price of an asset (be it financial or real) that pays no dividends (or convenience yield), $\delta = 0$. Assume that it follows a GBM. The risk-neutral version of this process is ($\lambda \equiv \rho\sigma\phi$):

$$ds = (\alpha - \lambda)Sdt + \sigma SdZ. \tag{4.1}$$

By now we know that in such a world the expected return on all the assets must equal the risk-free rate r, so for our particular asset $\alpha - \lambda = r$ (note that $\delta = 0$). Therefore, Eq. (4.1) can be stated equivalently as:

$$ds = rSdt + \sigma SdZ. \tag{4.2}$$

Hence the expected price at the end of a time interval Δt is:

$$E(S_t) = S_0 e^{r\Delta t}. \tag{4.3}$$

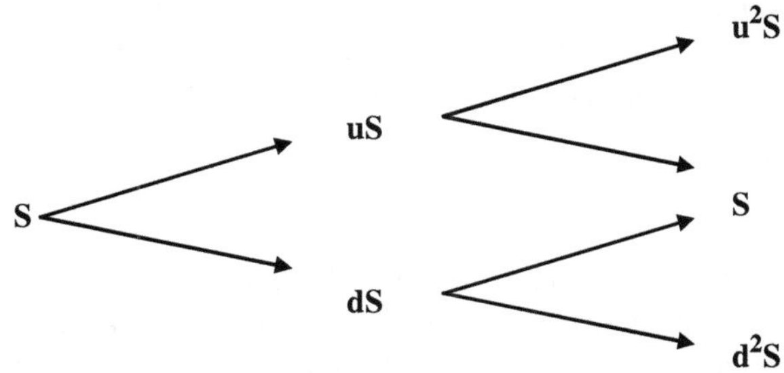

Fig. 4.1 Change in the price of the underlying asset under the binomial model with $u = 1/d$

And the variance approaches:

$$Var(S_t) = S_0^2 e^{2r\Delta t}\left(e^{\sigma^2 \Delta t} - 1\right) \approx \sigma^2 S_0^2 \Delta t. \tag{4.4}$$

Consider the valuation of a call option (worth C) on this asset. For simplicity, assume that the amount of the underlying asset controlled by the option contract is 1 unit. This movement of S in discrete time is going to be essentially the same as in the previous section. The time to the option's maturity is divided into a large number of short intervals of length Δt. Thus, by next period it will change by either a factor u (an up movement) or d (a down movement). At that time, the asset will either rise to uS or fall to ds; see Fig. 4.1. In general, $u > 1$ and $d < 1$. We assume that the probability of increase is p, and that of a decrease is $(1 - p)$.

4.2.1 Determining the Parameters of the Lattice

The lattice of the asset price aims to match as closely as possible the probability distribution in continuos time. The values of u, d, and p are determined in such a way that the former objective is accomplished. In particular, there must be a strict correspondence regarding the mean (or mathematical expectation) and also the variance over any short time interval Δt.

Concerning the former the following must hold:

$$Se^{r\Delta t} = puS + (1 - p)ds \leftrightarrow e^{r\Delta t} = pu + (1 - p)d. \tag{4.5}$$

As for the latter we get:

$$\begin{aligned} &\sigma^2 S^2 \Delta t = pu^2 S^2 + (1 - p)d^2 S^2 - [pu + (1 - p)d]^2 S^2 \\ &\leftrightarrow \sigma^2 \Delta t = pu^2 + (1 - p)d^2 - [pu + (1 - p)d]. \end{aligned} \tag{4.6}$$

We thus have two restrictions that the values of u, d, and p must satisfy. A third one that is frequently adopted is:

$$u = \frac{1}{d}. \tag{4.7}$$

Equation (4.5) directly gives the formula for the value of p. The other two restrictions, Eqs. (4.6) and (4.7), imply the following approximated values for u and d (provided Δt is small):

$$p = \frac{e^{r\Delta t} - d}{u - d}, u = e^{\sigma\sqrt{\Delta t}}, d = e^{-\sigma\sqrt{\Delta t}}. \tag{4.8}$$

These formulas allow build the whole lattice of the underlying asset price. In general, after a number i of time intervals (i.e. at date $i\Delta t$), there are $i + 1$ potential prices of the asset. This is because the lattice recombines: a rise followed by a fall leads to the same asset price that a fall followed by a rise (otherwise the number of nodes would be substantially higher). Each nodal price is given by the formula:

$$Su^j d^{i-j}, \tag{4.9}$$

where $j = 1, 2, \ldots, i$ stands for the number of upward movements.

Box 4.1 Risk neutrality and option valuation

Here we explain the role played by the assumption of risk neutrality in the valuation of options. To this end we consider the simplest case, namely the one-period formula. Assume that the initial asset price is S. In this case, when the option reaches maturity, the price will either increase to uS or decrease to dS. The value of the call option upon expiration will be either $C_u = \max(uS - K, 0)$ or $C_d = \max(dS - K, 0)$. See Fig. 4.2.

The basic idea that allows the valuation of option contracts is that we can form a particular portfolio, containing a specific number N of units of the underlying asset at its current price S while borrowing a suitable dollar amount B at the riskless interest rate r, which exactly replicates the future payoffs of the option in any state of nature. Since the call option and the replicating portfolio have the same future returns, to avoid any arbitrage opportunity they must trade at the same current price. Therefore, we can

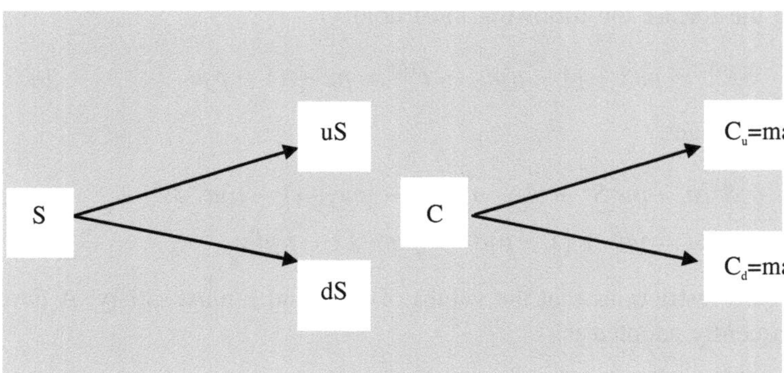

Fig. 4.2 Movement of the asset price and the option value in one period

value the option by computing the cost of forming this equivalent replicating portfolio:

$$C = NS - B. \tag{4.10}$$

The present value of this portfolio is $NS - B$. After one period, we have to pay back the amount borrowed initially plus accrued interest $(1 + r)B$ with certainty; in the meantime, our holdings of the underlying asset will be worth either NuS or NdS. Overall, next period the value of the replicating portfolio will be either $NuS - (1 + r)B$ or $NdS - (1 + r)B$. If this portfolio is to have the same payoffs that the option in any state of nature then:

$$NuS - (1 + r)B = C_u,$$
$$NdS - (1 + r)B = C_d.$$

Solving for the two unknowns N and B we get:

$$N = \frac{C_u - C_d}{uS - dS},$$
$$B = \frac{dC_u - uC_d}{(u - d)(1 + r)}.$$

Substituting these two expressions in Eq. (4.10) we have:

$$C = \frac{pC_u + (1 - p)C_d}{(1 + r)}, \text{ with } p \equiv \frac{(1 + r)S - ds}{uS - ds}. \tag{4.11}$$

The parameter p can be interpreted as a risk-neutral probability, i.e., the probability in a world where investors are risk neutral. To see this, note that the formula for C can be rearranged as $NS - C = B$. In words, forming a portfolio consisting in (a) the purchase of N units of the underlying asset, and (b) the sale of a call option would provide a dollar amount $(1 + r)B$ next period for certain, irrespective of whether the assets price rises or falls. The possibility to form this hedge renders the investors' risk profile irrelevant; i.e., any particular assumption in this regard will do. For convenience, we derive the value of the option in a risk-neutral world. In such a world, all the assets (stocks, options, ...) would earn the riskless interest rate. Therefore, the expected cash flows (weighted by risk-neutral probabilities) can be suitably discounted at the risk-free interest rate.

In this regard, let $R_u \equiv u - 1 = \frac{uS}{S} - 1$ denote the return if the asset price rises, and $R_d \equiv d - 1 = \frac{dS}{S} - 1$ the return if it falls. If the asset expected return in a risk-neutral world is to be equal to the risk-free interest rate then:

$$pR_u + (1 - p)R_d = r.$$

Solving for p we get:

$$p = \frac{r - R_d}{R_u - R_d} = \frac{(1+r) - d}{u - d}; \tag{4.12}$$

note that $(1 + r) \approx e^{r\Delta t}$ for Δt small; see Eq. (4.8). Similarly, in a risk-neutral world the option expected return must also equal the riskless interest rate:

$$\frac{pC_u + (1-p)C_d}{C} - 1 = r,$$

which results from the above formula for the option value C. See Trigeorgis (1996).

4.2.2 The Finite-Lived Option to Invest

Valuation of options starts from the end of the lattice (T) since their payoffs at that time are known. From these terminal payoffs, backward induction allows to derive the present value of the option.

Consider an American call option. The time to maturity is split into N subintervals each Δt long. Let f_{ij} denote the option value at node (i, j), i.e. at time $i\Delta t$ after j upward movements: $f_{ij} = Su^j d^{i-j}$. At time $T = N\Delta t$ the payoff from the option is:

$$f_{Nj} = \max\left(Su^j d^{N-j} - K, 0\right), \quad 0 \le j \le i. \tag{4.13}$$

In earlier periods, there is a probability p of moving up from node (i, j) at time $i\Delta t$ to node $(i+1, j+1)$ at time $(i+1)\Delta t$. And there is a probability $(1-p)$ of moving down from node (i, j) at time $i\Delta t$ to node $(i+1, j)$ at time $(i+1)\Delta t$. Neglecting for the time being the possibility of early exercise, risk-neutral valuation yields:

$$f_{ij} = e^{-r\Delta t}\left[pf_{i+1,\,j+1} + (1-p)f_{i+1,\,j}\right], \quad \text{for} \quad 0 \le i \le N-1, \quad 0 \le j \le i.$$

Note here that the value at one node is just the expected value resulting from the two nodes ahead of it, discounted over Δt at the riskless rate r.

Since the option is of the American type we have the opportunity to exercise early, so we must check at any time if this is indeed the best decision as compared to keeping the option alive for one more Δt. Therefore, this value must be set against the value of exercising immediately (the intrinsic value of the option):

$$f_{ij} = \max\left\{Su^j d^{i-j} - K,\ e^{-r\Delta t}\left[pf_{i+1,j+1} + (1-p)f_{i+1,j}\right]\right\}, \tag{4.14}$$

for $0 \le i \le N-1$, $0 \le j \le i$. Since we move from time T backward along the lattice, f_{ij} not only captures the value accruing from early exercise at time $i\Delta t$ but also that resulting from this feature in all subsequent dates. Thus, if we find it optimal to exercise the option at $i\Delta t$ we can be confident that we are not being naîve; quite the opposite, we come up with that conclusion after having pondered the best course of actions from then on. This is what dynamic programming is all about.

Note, however, that this approach is only valid when the current value of the option depends on the current values of the underlying variables but not on their past history (i.e. there is no path dependency, in the sense that the particular trail of swings that led to the current state is irrelevant).

4.2.3 Extensions

A dividend-paying underlying asset Now consider the case that the underlying asset pays a continuous dividend (or convenience yield) at the rate δ. Since the total return on the asset must amount to r in the risk-neutral world, the price return must be $(r-\delta)$. Equation (4.5) then becomes:

$$Se^{(r-\delta)\Delta t} = puS + (1-p)ds \leftrightarrow e^{(r-\delta)\Delta t} = pu + (1-p)d. \tag{4.15}$$

This affects the formula for the risk-neutral probability p, which now is:

$$p = \frac{e^{(r-\delta)\Delta t} - d}{u-d}, \tag{4.16}$$

but leaves the expressions for the parameters u and d unchanged. Since $r-\delta = \alpha - \lambda$ we can use them interchangeably: $e^{(r-\delta)\Delta t} = e^{(\alpha-\lambda)\Delta t}$. Further, when the underlying asset is a commodity, the (risk-neutral) rate of return $(r-\delta)$ can be estimated from the prices of futures contracts on that commodity.

There can be instances in which σ is very low and this leads to either p or $(1-p)$ being very small or even negative (in which case they make no sense). This can be avoided through the use of futures prices of the underlying asset.

Log transformation of the asset price Define $X = \ln S$; then applying Ito's Lemma:

$$dX = \left(\alpha - \rho\sigma\phi - \frac{1}{2}\sigma^2\right)dt + \sigma dZ = \hat{\alpha}dt + \sigma dZ. \tag{4.17}$$

It can be shown that:

$$\Delta X = \sigma\sqrt{\Delta t}, u = e^{\sigma\sqrt{\Delta t}}, \; d = e^{-\sigma\sqrt{\Delta t}}. \tag{4.18}$$

The risk-neutral probability of an upward movement is given by

$$p = \frac{1}{2} + \frac{\hat{\alpha}\sqrt{\Delta t}}{2\sigma}, \tag{4.19}$$

where $\hat{\alpha} \equiv \alpha - \rho\sigma\phi - \frac{1}{2}\sigma^2$ is a constant (i.e. it remains the same across all the nodes).

The binomial lattice using futures prices Now we are going to use the dynamics in futures prices to build the lattice. The value at time t of a futures contract with maturity at T is given by:

$$F(t, T) = S_t e^{(\alpha-\lambda)(T-t)}. \tag{4.20}$$

This equation shows that an asset whose spot price follows a GBM should trade on the futures market at quotes that increase (in absolute value terms) by greater amounts as times to maturity increase. This behaviour serves to identify underlying assets as candidates for modelling as a GBM.

The next step is to obtain the differential equation followed by the price of futures contracts over time:

$$dF = \left[\frac{\partial F}{\partial S}(\alpha - \lambda)S + \frac{\partial F}{\partial t} + \frac{1}{2}\frac{\partial^2 F}{\partial S^2}\sigma^2 S^2\right] dt + \frac{\partial F}{\partial S}\sigma S dZ.$$

Given that:

$$F_S \equiv \frac{\partial F}{\partial S} = e^{(\alpha-\lambda)(T-t)},\ F_{SS} \equiv \frac{\partial^2 F}{\partial S^2} = 0,\ F_t \equiv \frac{\partial F}{\partial t} = -(\alpha - \lambda)S_t e^{(\alpha-\lambda)(T-t)},$$

in the end we get:

$$dF = \sigma S_t e^{(\alpha-\lambda)(T-t)} dZ_t = \sigma F_t dZ_t. \tag{4.21}$$

In this case the drift has disappeared. This can be useful in numerical methods such as binomial lattices since the possibility of obtaining negative probabilities (in one-dimensional calculations) disappears. As shown here, in the risk-neutral world there can be no trend in futures prices; otherwise, positive returns could be obtained without investing anything.

Since futures prices on a GBM have no drift in the risk-neutral world ($\alpha - \lambda = 0$), the equations for the mean and the variance over a short Δt are:

$$E_0^Q(F_{\Delta t}) = F_0, Var(F_{\Delta t}) = F_0^2\left(e^{\sigma^2 \Delta t} - 1\right). \tag{4.22}$$

Thus, the parameters to be used in a (one-dimensional) binomial lattice must comply with the following:

1. for the mean:

$$F = pFu + (1 - p)Fd; \tag{4.23}$$

2. for the variance:

$$F^2\left(e^{\sigma^2\Delta t}-1\right)=pF^2u^2+(1-p)F^2d^2-F^2. \tag{4.24}$$

Hence the following can be deduced:

$$p=\frac{1-d}{u-d},\ e^{\sigma^2\Delta t}=pu^2+(1-p)d^2. \tag{4.25}$$

Parameters p, u, and d are independent of F. The probabilities are therefore constant throughout the tree. Since $u>1$ and $d<1, p>0$ is always satisfied. Similarly, $(1-p)=\frac{u-1}{u-d}>0$. Therefore, in this type of implementation the risk-neutral probabilities will always be positive.

The exact solution for the value of u can be found by solving the system of the last two equations. This can be proven to be:

$$u=\frac{e^{\sigma^2\Delta t}+1+\sqrt{\left(e^{\sigma^2\Delta t}+1\right)^2-4}}{2}, \tag{4.26}$$

a solution which is valid regardless of the size of Δt. Frequently, however, for short steps Δt, the next simplified solution is used:

$$u\cong e^{\sigma\sqrt{\Delta t}}.$$

In this binomial lattice, the trigger price corresponds to the price of the underlying asset S^* for which the immediate exercise value is greater than or equal to the continuation value. In the one-dimensional case, the optimal exercise price is a point on the border between the "invest region" and the "wait region".

The binomial lattice using log futures prices It is also possible to use the natural logarithms of futures prices for building binomial lattices.

The transformation $X=\ln F$ results in:

$$dX=-\frac{\sigma^2}{2}dt+\sigma dZ_t. \tag{4.27}$$

In this case, the equations to be satisfied when building the lattice are:

$$E(\Delta X)=p\Delta X+(1-p)\Delta X=-\frac{\sigma^2}{2}\Delta t, \tag{4.28}$$

$$E\left(\Delta X^2\right)=p\Delta X^2+(1-p)\Delta X^2=\sigma^2\Delta t+\frac{\sigma^4}{4}(\Delta t)^2\cong\sigma^2\Delta t. \tag{4.29}$$

This way the lattice recombines. From the last equation it is straightforward to derive $\Delta X=\sigma\sqrt{\Delta t}$. Therefore:

$$p = \frac{1}{2} - \frac{\sigma\sqrt{\Delta t}}{4}, (1-p) = \frac{1}{2} + \frac{\sigma\sqrt{\Delta t}}{4}. \tag{4.30}$$

In this case, the length of Δt chosen must be short enough to ensure that negative probabilities are not obtained.

4.2.4 Example 1: One Time Step Per Year

Consider the case with a crude time discretization: $T = 1$, and $\Delta t = 1$; i.e. the whole time horizon (a year) is approximated by a single time step. Although we can invest in a carbon-abating project either now or in a year's time, upon investment the savings (or sale revenues from spare allowances) will accrue over the next 20 years; $\tau_1 = 0$, $\tau_2 = 20$. The remaining parameters are as follows: $\alpha - \lambda = 0.05$, $r = 0.10$, $I = 100$, $\sigma = 0.40$ and $S_0 = 10$. First we assess the decision to invest at the end; then, at the beginning.

After one period, the emission allowance price can go either up or down:

$$S^+ \equiv uS = S_0 e^{\sigma\sqrt{\Delta t}} = 14.9182,\ S^- \equiv dS = S_0 e^{-\sigma\sqrt{\Delta t}} = 6.7032,$$

with probabilities:

$$p_u = \frac{e^{(r-\delta)\Delta t} - d}{u - d} = \frac{e^{(\alpha-\lambda)\Delta t} - e^{-\sigma\sqrt{\Delta t}}}{e^{\sigma\sqrt{\Delta t}} - e^{-\sigma\sqrt{\Delta t}}} = 0.4637,\ p_d = 1 - p_u = 0.5363,$$

respectively. In case the investment is undertaken after one year, we get the present value of a 20 year income:

$$V(S_T) = \frac{S_T}{\alpha - \lambda - r}\left[e^{(\alpha-\lambda-r)\tau_2} - e^{(\alpha-\lambda-r)\tau_1}\right] = 12.6424 S_T. \tag{4.31}$$

Thus, if we invest at the option expiration we receive one of these two payoffs:

$$V(14.9182) - 100 = 88.6026,\ V(6.7032) - 100 = -15.2554.$$

Note, however, that there is no obligation to go ahead with the investment irrespective of the outcome. Instead, we have an option to invest. So the terminal payoffs are either positive or zero (but never negative):

$$W^+ = \max(88.6026; 0) = 88.6026,\ W^- = \max(-15.2554; 0) = 0.$$

At the initial time there are two possibilities, namely whether to invest or to wait. In case of investing we receive the net present value:

$$NPV = V(10) - 100 = 126.4241 - 100 = 26.4241. \tag{4.32}$$

If, instead, we opt for waiting (thus keeping the option alive) we get the continuation value, i.e. the expectation of the terminal payoffs discounted to the present:

$$e^{-r\Delta t}(p_u W^+ + p_d W^-) = 37.1772. \tag{4.33}$$

Since $37.1772 > 26.4241$ it is clear that initially the optimal decision is to wait.

Now we can ask: given $S_0 = 10$, what is the threshold investment cost I^* below which it is optimal to invest immediately? This will be indeed the best course of action when the NPV surpasses the continuation value. Note, though, that both parties to the comparison depend on I. The reader can check that the two values become equal for $I^* = 64.7919$. Therefore, investment will take place at the initial time if $I < I^* = 64.7919$.

We can also ask: given $I = 100$, what is the trigger carbon price S^* above which it is optimal to invest immediately? As before, the two values to be compared depend on S. The critical price turns out to be $S^* = 15.4340$. Consequently, we will invest initially if $S > S^* = 15.4340$. This price level S^* separates the continuation (waiting) region from the (immediate) investment region.

This has been a very simple example for illustrative purposes. In practice it is not very correct to adopt $\Delta t = 1$, since the continuous-time process for S will hardly be adequately represented. If, for example, we use $\Delta t = 1/100$, i.e. we develop a binomial lattice with 100 steps in a year, we get an option value 35.6034. This is more accurate (and lower) than the above value 37.1772 with one step.

4.2.5 Example 2: One Hundred Time Steps Per Year

Assume now that the investment opportunity is available for 10 years, $T = 10$, but exercising this option at time t provides a perpetual income: $\tau_1 = 0, \tau_2 \to \infty$. The remaining parameter values are: $\alpha - \lambda = 0.05, r = 0.10, I = 100, \sigma = 0.40, S_0 = 10$, and $\Delta t = 1/100$. The gross PV of this infinite stream is:

$$V(S_t) = \frac{S_t}{r - (\alpha - \lambda)} = 20S_t. \tag{4.34}$$

In this case each time period comprises 100 time steps. Therefore, we build a lattice with $100 \times 10 = 1,000$ steps. The discretization is thus much finer, and the numerical result more accurate. The value of the option is 116.1728, while investing immediately yields $NPV = V(S_0) - I = 20 \times 10 - 100 = 100$. In sum, it is better to wait.

It is possible to compute the trigger cost for which the continuation value equals exactly the investment value at time 0 with $S_0 = 10$; it happens to be $I^* = 52.32$. Similarly, for $I = 100$ we determine the critical price $S^* = 19.11$.

4.2.6 Example 3: Convergence to the Perpetual Option

Assume now that the option to invest in this project is available forever, $T \to \infty$, but investment at time t provides a stream of cash flows over 20 years: $\tau_1 = 0, \tau_2 = 20$. The other parameters are: $\alpha - \lambda = 0.05, r = 0.10, I = 280, \sigma = 0.40, S_0 = 10$, and $\Delta t = 1/100$. The present value of the cash flows is:

$$V(S_t) = \frac{S_t}{\alpha - \lambda - r}\left[e^{(\alpha-\lambda-r)\tau_2} - e^{(\alpha-\lambda-r)\tau_1}\right] = 12.6424 S_t. \tag{4.35}$$

We know from Chap. 3 that the value of the perpetual option is: $H(S) = A_1 S^{\gamma_1}$. In this case, $\gamma_1 = 1.3211$, which does not depend upon the investment cost I; see Sect. 3.3.3. To compute A_1 we need the boundary conditions:

$$\text{Value} - \text{matching}: A_1 (S^*)^{\gamma_1} = 12.6424 S^* - 280.$$

$$\text{Smooth} - \text{pasting}: \gamma_1 A_1 (S^*)^{\gamma_1 - 1} = 12.6424.$$

From this system of two equations we compute $S^* = 91.1119$ and $A_1 = 2.2468$. Hence the perpetual option value is $A_1 (S_0)^{\gamma_1} = 47.0667$. Note that both values fall along the border between the continuation region and the investment region: $I = 3.0731 S^* = 3.0731 \times 91.1119 = 280$.

We are going to approximate the value of the perpetual option by means of a binomial lattice with 100 steps per year and a varying number of years to maturity (up to 100 years, i.e. 10,000 time steps in total). As the time to expiration increases the value of the finite-lived option approaches that of the perpetual option; this enables us to assess the accuracy or suitability of our code to compute binomial lattices. The numerical results appear in Table 4.1. We can check the good approximation to the infinite option with $T = 50$ and $T = 100$ years.

Results will be different if we keep $S_0 = 10$ but change the investment cost to $I = 30.73$. In this case, the value of the perpetual option (as seen in Chap. 3 with constant I) is 95.6927. Table 4.2 displays the results. Here also we see the convergence as the option's maturity rises.

Table 4.1 Convergence to the perpetual option under a GBM

Time to maturity (years)	Option value
0	0.0000
1	0.8123
5	18.4902
10	33.0342
25	45.2554
50	46.9779
100	47.0525
∞	47.0667

Table 4.2 Continuation value and investment value for different maturities

Time (years)	NPV	Wait	Max (NPV, wait)
50	95.6941	95.6771	95.6941
100	95.6941	95.6772	95.6941
∞	95.6927	95.6927	95.6927

4.2.7 Example 4: Decreasing Investment Cost (One Step Per Year)

This is the same as Example 1, but now investment cost decreases with the passage of time: $I_t = I_0 e^{\beta t} = 100e^{-0.25t}$. Therefore, at the option maturity we can end up with one the two following values:

$$V(14.9182) - 100e^{-0.25} = 91.0716,\ V(6.7032) - 100e^{-0.25} = -12.7864.$$

Consequently, the option's payoffs at each state are:

$$W^+ = \max(91.0716; 0) = 91.0716,\ W^- = \max(-12.7864; 0) = 0.$$

At the initial time, investing immediately delivers a net present value of:

$$NPV = V(10) - 100 = 126.4241 - 100 = 26.4241. \qquad (4.36)$$

The continuation value amounts to:

$$e^{-r\Delta t}(p_u W^+ + p_d W^-) = 38.2131. \qquad (4.37)$$

Therefore, it is not optimal to invest initially. This is partly due to the decreasing character of the investment cost.

4.2.8 Example 5: Decreasing Investment Cost (One Hundred Steps Per Year)

This is the same as Example 4 above, i.e. $I_t = I_0 e^{\beta t} = 100e^{-0.25t}$, but now the time to maturity (one year) is divided into 100 steps, thus $\Delta t = 1/100$. In this case we get a value of the option 37.1634. This is less than 38.2131 before.

4.2.9 Example 6: Convergence to Perpetual Option (Decreasing Investment Cost)

Again we consider an infinite-lived option to invest and the investment provides cash flows over 20 years. Also: $\alpha - \lambda = 0.05, r = 0.10, I = 280, \sigma = 0.40, S_0 = 10$, and $\Delta t = 1/100$. The investment cost is anticipated to decrease over time according to a deterministic pattern. We define the ratio $x \equiv S/I$. The option value can be written as: $h(x) = A_1 x^{\gamma_1}$. We consider that the investment provides cash flows over 20 years; the option to invest is available either for 50 years, 100 years or forever. The results appear in Table 4.3.

As shown in Chap. 3, for the perpetual option we get $\gamma_1 = 1.2816$ and $x^* = 0.3599$; see Sect. 3.3.3. Hence we compute $A_1 = \frac{12.6424(x^*)^{1-\gamma_1}}{\gamma_1} = 13.15359$. Therefore, the value of the option is:

$$H(S, I) = Ih(x) = A_1 I x^{\gamma_1} = 13.15359 \times 280 \times \left(\frac{10}{280}\right)^{1.2816} = 51.4595.$$

For $I = 280$, from $x^* = 0.3599$ we would derive $S^* = 100.7858$, and the option value would be:

$$H(S^*, I) = 13.15359 \times 280 \times \left(\frac{100.7858}{280}\right)^{1.2816} = 994.175. \tag{4.38}$$

This option value will coincide at the initial time with the net present value:

$$NPV = 12.6424x^* - I = 12.6424 \times 100.7858 - 280 = 994.175. \tag{4.39}$$

4.3 The Finite-Lived Option to Invest Under the IGBM

In this case F satisfies the partial differential Eq. (3.45):

$$\frac{1}{2}\sigma^2 S^2 \frac{\partial^2 F}{\partial S^2} + [k(S_m - S) - \rho\sigma\phi S]\frac{\partial F}{\partial S} + \frac{\partial F}{\partial t} - rF = 0,$$

Table 4.3 Convergence to the perpetual option

Time to option expiration (years)	Option value
10	38.0593
50	51.4168
100	51.4471
∞	51.4595

which must be solved by means of numerical procedures. Given the American type of the options here involved and the low number of sources of uncertainty, the binomial lattice approach is used.

Following this approach, the time horizon T is subdivided into n steps, each of size $\Delta t = T/n$. Starting from an initial value S_O, at time i, after j positive increments, the price of the underlying asset (say, oil) is given by $S_O u^j d^{i-j}$, where $d = 1/u$.

Consider an asset whose risk-neutral behavior follows Eq. (3.12)

$$dS = [k(S_m - S) - \rho\sigma\phi S]dt + \sigma S dZ.$$

This can also be written as:

$$dS = \left(\frac{k(S_m - S)}{S} - \rho\sigma\phi\right) S dt + \sigma S dZ_t = \mu S dt + \sigma S dZ.$$

Since it is usually easier to work with the processes for the natural logarithms of asset prices, we carry out the following transformation: $X = \ln S$. Thus, $X_S = 1/S$, $X_{SS} = -1/S^2$, and $X_t = 0$; by Ito's Lemma:

$$dX = \left[\frac{k(S_m - S)}{S} - \rho\sigma\phi - \frac{1}{2}\sigma^2\right] dt + \sigma dZ = \hat{\mu} dt + \sigma dZ, \tag{4.40}$$

where $\hat{\mu}$ depends at each moment on current price S.

Following Euler–Maruyama's discretization, the probabilities of upward and downward movements must satisfy three conditions:

(a) $p_u + p_d = 1$.

(b) $E(\Delta X) = p_u \Delta X - p_d \Delta X = \left[\frac{k(S_m - S)}{S} - \rho\sigma\phi - \frac{1}{2}\sigma^2\right] \Delta t = \hat{\mu}\Delta t$. The aim is to equate the first moment of the binomial lattice ($p_u \Delta X - p_d \Delta X$) to the first moment of the risk-neutral underlying variable ($\hat{\mu}\Delta t$).

(c) $E(\Delta X^2) = p_u \Delta X^2 + p_d \Delta X^2 = \sigma^2 \Delta t + \hat{\mu}^2 (\Delta t)^2$. In this case the equality refers to the second moments. For small values of Δt, we have $E(\Delta X^2) \approx \sigma^2 \Delta t$.

From (a) and (b) we obtain the probabilities, which can be different at each point of the lattice (because $\hat{\mu}$ depends on S, which varies from node to node):

$$p_u = \frac{1}{2} + \frac{\hat{\mu}\Delta t}{2\Delta X}.$$

From (c) there results:

$$\Delta X = \sigma\sqrt{\Delta t} \rightarrow u = e^{\sigma\sqrt{\Delta t}},\ d = e^{-\sigma\sqrt{\Delta t}}. \tag{4.41}$$

The probability of an upward movement at node (i, j) is

$$p_u(i, j) = \frac{1}{2} + \frac{\hat{\mu}(i, j)\sqrt{\Delta t}}{2\sigma}, \tag{4.42}$$

where:

$$\hat{\mu}(i,j) \equiv \frac{k\left(S_m - \hat{S}(i,j)\right)}{\hat{S}(i,j)} - \rho\sigma\phi - \frac{1}{2}\sigma^2. \tag{4.43}$$

4.3.1 Example 7: One Time Step Per Year

Now we consider an investment in a facility which consumes one barrel of oil every year. The investment provides a flow of income or revenues whose present value amounts to $I = 1,500$, but entails the purchase of a barrel per year over 20 years: $\tau_1 = 0, \tau_2 = 20$,The option is available for one year ($T = 1$) and we take one time step per year ($\Delta t = 1$). Other parameter values are: $S_0 = 110$, $\frac{kS_m}{k+\lambda} = 90$, $r = 0.02$, $k + \lambda = 0.30$, and $\sigma = 0.25$.

After one period, $\Delta t = 1$, oil price can rise to $S_0 e^{\sigma\sqrt{\Delta t}} = 141.2428$ or fall to $S_0 e^{-\sigma\sqrt{\Delta t}} = 85.6681$. The transformation $X = \ln S$ results in $\Delta X = \sigma\sqrt{\Delta t}$ and a probability $p_u = \frac{1}{2} + \frac{\hat{\mu}\sqrt{\Delta t}}{2\sigma} = 0.3284$, where $\hat{\mu}$ is:

$$\hat{\mu} \equiv \frac{\frac{kS_m}{k+\lambda}(k+\lambda) - (k+\lambda)S_0}{S_0} - \frac{1}{2}\sigma^2 = \frac{k(S_m - S_0)}{S_0} - \lambda - \frac{1}{2}\sigma^2 = -0.0858.$$

The probability of a downward change is: $p_d = 1 - p_u = 0.6716$.

In case of investing after one year we get $I = 1,500$ less the present value of a yearly rent over 20 years [see Eq. (3.15)]:

$$NPV(S_1) = 1,500 - \left[\frac{S_1 - \frac{kS_m}{k+\lambda}}{k+\lambda+r}\left(e^{-(k+\lambda+r)\tau_1} - e^{-(k+\lambda+r)\tau_2}\right) + \frac{kS_m}{r(k+\lambda)}\left(e^{-r\tau_1} - e^{-r\tau_2}\right)\right]. \tag{4.44}$$

For $S_1 = 141.2428$ we derive $NPV(141.2428) = -143.4275$. Thus, the option payoff in this state is $W^+ = \max[-143.4275; 0] = 0$. If, instead, we have $S_1 = 85.6681$ we derive $NPV(85.6681) = 29.9549$, so the option payoff in this state is $W^- = \max[29.9549; 0] = 29.9549$. At time 0 the continuation value is $e^{-r\Delta t}(p_u W^+ + p_d W^-) = 19.7191$. Against this alternative, the value of investing immediately (at $t = 0$) is $NPV(110) = -45.9559$. Therefore, there is no initial investment, and the value of the option is 19.7191.

4.3.2 Example 8: One Hundred Time Steps Per Year

The numerical data remain the same as in Example 9 with only one exception, namely $\Delta t = 1/100$ now. We thus build a binomial lattice with $100 \times 1 = 100$ steps, whose numerical result is more accurate. The value of the option is 15.3954 (lower than in the above example).

4.3.3 Example 9: Convergence to the Perpetual Option

Again $\Delta t = 1/100$, while $I = 1,500, \tau_1 = 0, \tau_2 = 20, S_0 = 110, \frac{kS_m}{k+\lambda} = 90$, $r = 0.02$, $k + \lambda = 0.30$, and $\sigma = 0.25$. The option is available for a number of different years to check the sensitivity of the value to the option maturity. Table 4.4 displays the resulting option values. As the time to expiration lenghthens we approach the amount provided by the analytic solution to the perpetual option (as seen in Chap. 3). For $S_0 = 110$, the latter is 98.5127.

4.4 Bi-dimensional Binomial Lattices

It is possible to develop binomial lattices when there is more than one underlying asset: the lattice unfolds into several dimensions. Again, it is key that the parameters take on values such that each state variable has an expected growth rate and standard deviation that are correct in the risk-neutral world. The correlation coefficient between any pair of state variables must also be correct. In practice, this approach is subject to the curse of dimensionality, and it becomes burdensome for more than two variables. We thus restrict ourselves to show the two-dimensional lattice.

Table 4.4 Convergence to the perpetual option

Maturity (years)	Option value	Maturity (years)	Option value
0	0	25	91.7670
1	15.3954	50	97.7397
5	50.4049	100	98.4798
10	70.6181	150	98.4916
15	81.5182	200	98.4918
20	87.8896	∞	98.5127

4.4.1 Example 10: Two GBMs

Clewlow and Strickland (1998) show a multidimensional lattice with two assets that follow correlated GBM's. Once more we adopt the parameter values: $T = 1$, $\Delta t = 1$, $r = 0.10, \tau_1 = 0$, and $\tau_2 = 20$. Regarding the two stochastic processes we adopt: $S_0 = 10, \alpha_S - \lambda_S = 0.05, \sigma_S = 0.40, I_0 = 100, \alpha_I - \lambda_I = 0.03, \sigma_I = 0.20$, and $\rho = 0.50$. After one period, $\Delta t = 1$, the asset price can rise to $S_0 e^{\sigma_S \sqrt{\Delta t}} = 14.9182$ or fall to $S_0 e^{-\sigma_S \sqrt{\Delta t}} = 6.7032$. Similarly, the investment cost can increase to $I_0 e^{\sigma_I \sqrt{\Delta t}} = 122.1403$ or decrease to $I_0 e^{-\sigma_I \sqrt{\Delta t}} = 81.8731$. Therefore, at $t = 1$ there are four possible states with their corresponding option payoffs and probabilities; see Table 4.5.

The transformations $X_1 = \ln S$ and $X_2 = \ln I$ result in $\Delta X_1 = \sigma_S \sqrt{\Delta t} = 0.40$ and $\Delta X_2 = \sigma_I \sqrt{\Delta t} = 0.20$. On the other hand:

$$\hat{\alpha}_1 = \alpha_S - \lambda_S - \frac{1}{2}\sigma_S^2 = -0.03, \hat{\alpha}_2 = \alpha_I - \lambda_I - \frac{1}{2}\sigma_I^2 = 0.01. \tag{4.45}$$

The probability of each state is given by the following probabilities:

$$p_{uu} = \frac{\Delta X_1 \Delta X_2 + \Delta X_2 \hat{\alpha}_1 \Delta t + \Delta X_1 \hat{\alpha}_2 \Delta t + \rho \sigma_S \sigma_I \Delta t}{4 \Delta X_1 \Delta X_2} = 0.36875, \tag{4.46}$$

$$p_{ud} = \frac{\Delta X_1 \Delta X_2 + \Delta X_2 \hat{\alpha}_1 \Delta t - \Delta X_1 \hat{\alpha}_2 \Delta t - \rho \sigma_S \sigma_I \Delta t}{4 \Delta X_1 \Delta X_2} = 0.09375, \tag{4.47}$$

$$p_{du} = \frac{\Delta X_1 \Delta X_2 - \Delta X_2 \hat{\alpha}_1 \Delta t + \Delta X_1 \hat{\alpha}_2 \Delta t - \rho \sigma_S \sigma_I \Delta t}{4 \Delta X_1 \Delta X_2} = 0.15625, \tag{4.48}$$

$$p_{dd} = \frac{\Delta X_1 \Delta X_2 - \Delta X_2 \hat{\alpha}_1 \Delta t - \Delta X_1 \hat{\alpha}_2 \Delta t + \rho \sigma_S \sigma_I \Delta t}{4 \Delta X_1 \Delta X_2} = 0.38125. \tag{4.49}$$

If, after one year, we invest with $S_1 = 14.9182$ and $I_1 = 122.1403$ we get the present value of a 20 year income less the investment cost:

$$NPV(S_1, I_1) = \frac{S_1}{\alpha_S - \lambda_S - r} \left[e^{(\alpha_S - \lambda_S - r)\tau_2} - e^{(\alpha_S - \lambda_S - r)\tau_1} \right] - I_1. \tag{4.50}$$

Table 4.5 Terminal payoffs and probabilities under two GBMs

State	S_1	I_1	Probability	$\max(V(S_1) - I_1; 0)$
uu	14.9182	122.1403	0.36875	$W^{++} = 66.4623$
ud	14.9182	81.8731	0.09375	$W^{+-} = 106.729$
du	6.7032	122.1403	0.15625	$W^{-+} = 0.000$
dd	6.7032	81.8731	0.38125	$W^{--} = 2.8715$

In particular, $NPV(14.9182, 122.1403) = 66.4623$. In the final nodes we will only invest if the outcome of doing so is $NPV(S_1, I_1) > 0$.

Now, the decision at the initial time is whether to invest or not; this decision comes down to comparing two values:

$$\max\left[NPV(S_0, I_0), e^{-r\Delta t}(p_{uu}W^{++} + p_{ud}W^{+-} + p_{du}W^{-+} + p_{dd}W^{--})\right] = 32.22$$

The continuation value (32.22) is higher than the investment value (26.4241), so the investment option will not be exercised initially.

4.4.2 Example 11: Two GBMs; Approximation to the Perpetual Option

Once more we consider an infinite-lived option to invest. The investment entails a cost $I = 100$, and provides cash flows over 20 years. Also: $\alpha - \lambda = 0.05$, $r = 0.10$, $\sigma = 0.40$, and $S_0 = 10$. Let x denote the ratio $x \equiv S/I$. The option to invest is approached numerically assuming that it is available for 50 years or alternatively 100 years; in both cases $\Delta t = 1/12$. The results appear in Table 4.6.

For the perpetual option we compute $\gamma_1 = 1.4637$ and $x^* = 0.249673$; see Sect. 3.3.4. Hence we get $A_1 = \frac{12.6424(x^*)^{1-\gamma_1}}{\gamma_1} = 16.43703$. The option value is:

$$H(S, I) = A_1 I x^{\gamma_1} = 16.43703 \times 100 \times \left(\frac{10}{100}\right)^{1.4637} = 56.5069. \tag{4.51}$$

From $x^* = 0.249673$, for $S^* = 10$ we derive $I^* = 40.05242$. For these particular figures, the continuation value (of the perpetual option) and the NPV are equal.

4.4.3 Two IGBMs

Consider two assets whose prices are governed by the following risk-neutral processes:

$$ds_1 = [k_1(S_{m_1} - S_1) - \rho_1\sigma_1\phi S_1]dt + \sigma_1 S_1 dZ_1, \tag{4.52}$$

Table 4.6 Convergence to the perpetual option

Time to option expiration (years)	Option value
10	50.8762
50	56.2657
100	56.3259
∞	56.5069

$$ds_2 = [k_2(S_{m_2} - S_2) - \rho_2\sigma_2\phi S_2]dt + \sigma_2 S_2 dZ_2, \tag{4.53}$$

$$dZ_1 dZ_2 = \rho_{12}, \tag{4.54}$$

where ρ_1 and ρ_2 denote the correlations of their respective returns with those of the market portfolio. Adopting the transformations $X_1 = \ln S_1, X_2 = \ln S_2$, and applying Ito's Lemma:

$$dX_1 = \left[\frac{k_1(S_{m_1} - S_1)}{S_1} - \rho_1\sigma_1\phi - \frac{1}{2}\sigma_1^2\right]dt + \sigma_1 dZ_1 = \hat{\mu}_1 dt + \sigma_1 dZ_1,$$

$$dX_2 = \left[\frac{k_2(S_{m_2} - S_2)}{S_2} - \rho_2\sigma_2\phi - \frac{1}{2}\sigma_2^2\right]dt + \sigma_2 dZ_2 = \hat{\mu}_2 dt + \sigma_2 dZ_2.$$

Now it is necessary to solve a system of six equations:

(a) $\boldsymbol{p_{uu} + p_{ud} + p_{du} + p_{dd} = 1}$. The probabilities must sum to one.
(b) $\boldsymbol{p_{uu} + p_{ud} + p_{du} + p_{dd} = 1}$. This is the expected value of the increment in X_1.
(c) $\boldsymbol{E(\Delta X_1^2) = (p_{uu} + p_{ud})\Delta X_1^2 + (p_{du} + p_{dd})\Delta X_1^2 = \sigma_1^2\Delta t + \hat{\mu}_1^2\Delta t^2}$. This refers to the second non-central moment of the increment in X_1.
(d) $\boldsymbol{E(\Delta X_2) = (p_{uu} + p_{du})\Delta X_2 - (p_{ud} + p_{dd})\Delta X_2 = \hat{\mu}_2\Delta t}$. This is the expected value of the increment in X_2.
(e) $\boldsymbol{E(\Delta X_2^2) = (p_{uu} + p_{du})\Delta X_2^2 + (p_{ud} + p_{dd})\Delta X_2^2 = \sigma_2^2\Delta t + \hat{\mu}_2^2\Delta t^2}$. This concerns the second non-central moment of the increment in X_2.
(f) $\boldsymbol{E(\Delta X_1\Delta X_2) = (p_{uu} - p_{ud} - p_{du} + p_{dd})\Delta X_1\Delta X_2 = \rho\sigma_1\sigma_2\Delta t + \hat{\mu}_1\hat{\mu}_2\Delta t^2}$. This is the expected value of the cross product ΔX_1, ΔX_2, which amounts to satisfying the correlation condition.

The solution to this system of equations, ignoring the terms in Δt^2, is:

$$\Delta X_1 = \sigma_1\sqrt{\Delta t}, \Delta X_2 = \sigma_2\sqrt{\Delta t}, \tag{4.55}$$

$$p_{uu} = \frac{\Delta X_1\Delta X_2 + \Delta X_2\hat{\mu}_1\Delta t + \Delta X_1\hat{\mu}_2\Delta t + \rho\sigma_1\sigma_2\Delta t}{4\Delta X_1\Delta X_2}, \tag{4.56}$$

$$p_{ud} = \frac{\Delta X_1\Delta X_2 + \Delta X_2\hat{\mu}_1\Delta t - \Delta X_1\hat{\mu}_2\Delta t - \rho\sigma_1\sigma_2\Delta t}{4\Delta X_1\Delta X_2}, \tag{4.57}$$

$$p_{du} = \frac{\Delta X_1\Delta X_2 - \Delta X_2\hat{\mu}_1\Delta t + \Delta X_1\hat{\mu}_2\Delta t - \rho\sigma_1\sigma_2\Delta t}{4\Delta X_1\Delta X_2}, \tag{4.58}$$

$$p_{dd} = \frac{\Delta X_1\Delta X_2 - \Delta X_2\hat{\mu}_1\Delta t - \Delta X_1\hat{\mu}_2\Delta t + \rho\sigma_1\sigma_2\Delta t}{4\Delta X_1\Delta X_2}. \tag{4.59}$$

In the above expressions, p_{ud} stands for the risk-neutral probability of an upward movement in asset 1's price and a simultaneous downward movement in asset 2's price at a certain node; similarly for the probabilities p_{uu}, p_{du}, and p_{dd}.

The branches of the lattice have been forced to recombine by taking constant increments ΔX_1 and ΔX_2 once the step size Δt has been chosen; thus, it is easier to implement the model in a computer program. However, the probabilities change from one node to another by depending on $\hat{\mu}_1$ and $\hat{\mu}_2$. Besides, it is necessary that at any time the four probabilities take on values between zero and one.

4.4.4 Example 12: Two IGBMs, One Step Per Year

We turn back to the numerical parameters for S in Example 7: $T = 1, \Delta t = 1$, $\tau_1 = 0, \tau_2 = 20, \tau_2 = 20, \frac{k_S S_m}{k_S + \lambda_S} = 90$, $r = 0.02$, $k_S + \lambda_S = 0.30$, and $\sigma_S = 0.25$. Now, however, the value of the revenues is also stochastic: $I_0 = 1,500, \frac{k_I I_m}{k_I + \lambda_I} = 1,500$, $k_I + \lambda_I = 0.30$, and $\sigma_I = 0.20$. The correlation coefficient between both processes is $\rho = 0.50$.

After one period, $\Delta t = 1$, oil price can rise to $S_0 e^{\sigma_S \sqrt{\Delta t}} = 141.2428$ or fall to $S_0 e^{-\sigma_S \sqrt{\Delta t}} = 85.6681$. Similarly the income (net from all other expenses) can increase to $I_0 e^{\sigma_I \sqrt{\Delta t}} = 1,832.1041$ or decrease to $I_0 e^{-\sigma_I \sqrt{\Delta t}} = 1,228.0961$.

The transformations $X_1 = \ln S$ and $X_2 = \ln I$ result in $\Delta X_1 = 0.25$ and $\Delta X_2 = 0.20$. In this case we also have:

$$\hat{\mu}_1 \equiv \frac{\frac{k_S S_m}{k_S + \lambda_S}(k_S + \lambda_S) - (k_S + \lambda_S) S_0}{S_0} - \frac{1}{2}\sigma_S^2 = -0.0858, \tag{4.60}$$

$$\hat{\mu}_2 \equiv \frac{\frac{k_I I_m}{k_I + \lambda_I}(k_I + \lambda_I) - (k_I + \lambda_I) I_0}{I_0} - \frac{1}{2}\sigma_I^2 = -0.02. \tag{4.61}$$

A simultaneous rise in both S and I is just one possibility, whose probability is:

$$p_{uu} = \frac{\Delta X_1 \Delta X_2 + \Delta X_2 \hat{\mu}_1 \Delta t + \Delta X_1 \hat{\mu}_2 \Delta t + \rho \sigma_S \sigma_I \Delta t}{4 \Delta X_1 \Delta X_2} = 0.2642. \tag{4.62}$$

At $t = 1$ there are three other possible states with their corresponding payoffs and probabilities: $p_{ud} = 0.0642$, $p_{du} = 0.1858$, and $p_{dd} = 0.4858$.

At the option's maturity, we get I_1 less the present value of a yearly rent over 20 years:

$$NPV(S_1, I_1) = I_1 - \left[\frac{S_1 - \frac{k_S S_m}{k_S + \lambda_S}}{k_S + \lambda_S + r} \left(e^{-(k_S + \lambda_S + r)\tau_1} - e^{-(k_S + \lambda_S + r)\tau_2} \right) + \frac{k S_m}{r(k + \lambda)} \left(e^{-r\tau_1} - e^{-r\tau_2} \right) \right]. \tag{4.63}$$

For $S_1 = 141.2428$ and $I_1 = 1832.1041$ we derive $NPV = 188.6767$. Thus the option payoff in this state is $W^{++} = \max[188.6767; 0] = 188.6767$. If we have $S_1 = 141.2428$ but, instead, $I_1 = 1228.0961$ then we derive $NPV = -415.3313$;

Table 4.7 Terminal payoffs and probabilities under two IGBMs

State	S_1	I_1	Probability	$\max(V(S_1) - I_1; 0)$
uu	141.2428	1,832.1041	0.2642	$W^{++} = 188.676$
ud	141.2428	1,228.0961	0.0642	$W^{+-} = 0$
du	85.6681	1,832.1041	0.1858	$W^{-+} = 362.059$
dd	85.6681	1,228.0961	0.4858	$W^{--} = 0$

therefore the option payoff in this state is $W^{+-} = \max[-415.3313; 0] = 0$. The other states appear in Table 4.7.

At $t = 0$ the continuation value is $e^{-r\Delta t}(p_{uu}W^{++} + p_{ud}W^{+-} + p_{du}W^{-+} + p_{dd}W^{--}) = 114.7991$. Against this alternative, the value of investing immediately (at $t = 0$) is $NPV(110, 1,500) = -45.9559$. Since the NPV is negative, it is optimal to exploit the possibility to wait.

4.4.5 Example 13: Two IGBMs with One Thousand Steps

The numerical parameter values are the same as in Example 12 with two exceptions: $T = 10$, and $\Delta t = 1/100$. Again, both S and I are stochastic and show mean reversion. The investment option is available over 10 years; and, upon investment, a stream of fuel payments takes place over 20 years. The option value turns out to be 280.8616, while the NPV amounts to -45.9559. The obvious choice is to wait.

4.4.6 One GBM and One IGBM

We have two risk-neutral stochastic processes, one for each source of risk e.g. the price of a carbon allowance or that of electricity. For the natural logarithm of the carbon allowance price:

$$dX_t = \left(\alpha_c - \frac{\sigma_c^2}{2} - \lambda_c\right)dt + \sigma_c dZ_t^c = \hat{\mu}_1 dt + \sigma_c dZ_t^c. \tag{4.64}$$

For the electricity price, assuming:

$$dE_t = \left[k_e\left(S_m^e - E_t\right) - \lambda_e E_t\right]dt + \sigma_e dZ_t^e,$$

the logarithmic transformation $Y_t \equiv \ln E_t$ yields:

$$dY_t = \left(\frac{k_e\left(S_m^e - E_t\right)}{E_t} - \lambda_e - \frac{\sigma_e^2}{2}\right)dt + \sigma_e dZ_t^e = \hat{\mu}_2 dt + \sigma_e dZ_t^e, \tag{4.65}$$

with:

$$dZ_t^c dZ_t^e = \rho dt. \tag{4.66}$$

There are four probabilities in the corresponding two-dimensional binomial lattice and, if we want the branches to recombine, two incremental values (ΔX and ΔY). At any time the four probabilities must take on values between zero and one, and add to one. Besides, they must be consistent with means, variances and correlations. So there are six restrictions to be satisfied. It can be shown that the solution is:

$$\Delta X = \sigma_c \sqrt{\Delta t}, \Delta Y = \sigma_e \sqrt{\Delta t}, \tag{4.67}$$

$$p_{uu} = \frac{\Delta X \Delta Y + \Delta Y \hat{\mu}_1 \Delta t + \Delta X \hat{\mu}_2 \Delta t + \rho \sigma_c \sigma_e \Delta t}{4 \Delta X \Delta Y}, \tag{4.68}$$

$$p_{ud} = \frac{\Delta X \Delta Y + \Delta Y \hat{\mu}_1 \Delta t - \Delta X \hat{\mu}_2 \Delta t - \rho \sigma_c \sigma_e \Delta t}{4 \Delta X \Delta Y}, \tag{4.69}$$

$$p_{du} = \frac{\Delta X \Delta Y - \Delta Y \hat{\mu}_1 \Delta t + \Delta X \hat{\mu}_2 \Delta t - \rho \sigma_c \sigma_e \Delta t}{4 \Delta X \Delta Y}, \tag{4.70}$$

$$p_{dd} = \frac{\Delta X \Delta Y - \Delta Y \hat{\mu}_1 \Delta t - \Delta X \hat{\mu}_2 \Delta t + \rho \sigma_c \sigma_e \Delta t}{4 \Delta X \Delta Y}. \tag{4.71}$$

Note that the drift rate for the allowance price ($\hat{\mu}_1$) is a constant, whereas that for the electricity price ($\hat{\mu}_2$) depends on E_t which changes from node to node. Consequently the four probabilities change from a node to the next. The two subscripts (u, d) refer to the allowance price and the electricity price, respectively.

4.5 Trinomial Lattice with Mean Reversion

The investment time horizon T is subdivided in n steps, each of size $\Delta t = T/n$. Starting from an initial electricity price S_0, in a trinomial lattice one of three possibilities will take place: either the price jumps up (by a factor u to S^+), remains the same ($S^=$), or jumps down (by a factor d to S^-). At time i, after j positive increments, the price is given by $S_0 u^j d^{i-j}$, where $d = 1/u$.

Consider an asset whose price follows the risk-neutral process:

$$dS_t = [k(S_m - S_t) - \lambda S_t]dt + \sigma S_t dZ.$$

This can also be written as:

$$dS_t = \left(\frac{k(S_m - S_t)}{S_t} - \lambda\right) S_t dt + \sigma S_t dZ_t.$$

It is usually easier to work with the processes for the natural logarithms of asset prices. Consequently we undertake a logarithmic transformation: $X = \ln S$. Thus, $X_S = 1/S$, $X_{SS} = -1/S^2$, and $X_t = 0$; by Ito's Lemma:

$$dX = \left[\frac{k(S_m - S_t)}{S_t} - \lambda - \frac{1}{2}\sigma^2\right]dt + \sigma dZ = \hat{\mu}dt + \sigma dZ,$$

where $\hat{\mu} \equiv \frac{k(S_m - S_t)}{S_t} - \lambda - \frac{1}{2}\sigma^2$ depends at each moment on the asset value S_t (so strict notation would read $\hat{\mu}(t)$).

In a trinomial lattice, there are three probabilities p_u, p_m, and p_d associated with a rise, maintenance, and a fall in the price of the asset. Following Euler–Maruyama's discretization, these probabilities must satisfy three conditions:

(a) $p_u + p_m + p_d = 1$.
(b) $E(\Delta X) = p_u \Delta X + p_m 0 - p_d \Delta X = \hat{\mu}\Delta t$. The aim is to equate the first moment of the binomial lattice ($p_u \Delta X - p_d \Delta X$) to the first moment of the risk-neutral underlying variable ($\hat{\mu}\Delta t$).
(c) $E(\Delta X^2) = p_u \Delta X^2 + p_m 0 + p_d \Delta X^2 = \sigma^2 \Delta t + \hat{\mu}^2 (\Delta t)^2$. In this case the equality refers to the second moments. For small values of Δt, we have $E(\Delta X^2) \approx \sigma^2 \Delta t$.

Solving the system for the three probabilities (Hull and White 1994) we get:

$$p_u = \frac{1}{2}\left[\frac{\sigma^2 \Delta t + \hat{\mu}^2 (\Delta t)^2}{(\Delta X)^2} + \frac{\hat{\mu}\Delta t}{2\Delta X}\right], \tag{4.72}$$

$$p_m = 1 - \frac{\sigma^2 \Delta t + \hat{\mu}^2 (\Delta t)^2}{(\Delta X)^2}, \tag{4.73}$$

$$p_d = \frac{1}{2}\left[\frac{\sigma^2 \Delta t + \hat{\mu}^2 (\Delta t)^2}{(\Delta X)^2} - \frac{\hat{\mu}\Delta t}{2\Delta X}\right]. \tag{4.74}$$

The particular values depend on $\hat{\mu}$, which changes from one node to the next. Specifically:

$$\hat{\mu}(i,j) \equiv \frac{k(S_m - S_t(i,j))}{S_t(i,j)} - \lambda - \frac{1}{2}\sigma^2. \tag{4.75}$$

So the three probabilities also change from one node to the next. In a trinomial lattice after n periods we have $2n + 1$ final nodes. This holds true irrespective of the initial commodity price.

In a trinomial lattice, as compared to a binomial one, there is an additional degree of freedom (there is a third possibility -the price to stay the same- while the three conditions remain unchanged). Thus, we can choose the size of the time step Δt; it is particularly convenient to choose its value in such a way that negative probabilities are avoided. Given that a trinomial lattice is basically an explicit difference scheme (Clewlow and Strickland 1998), convergence and stability reasons suggest to adopt $\Delta X = \sigma\sqrt{3\Delta t}$ (Hull and White 1994). In this case:

$$p_u = \frac{1}{6} + \frac{M^2 + M}{2}, \quad M \equiv \frac{\hat{\mu}\Delta t}{\sigma\sqrt{3\Delta t}},$$

$$p_m = \frac{2}{3} - M^2,$$

$$p_d = \frac{1}{6} + \frac{M^2 - M}{2}.$$

When, in principle, $p_u < 0$, the three possibilities that we choose for the asset price are: stay unchanged, fall by $-\Delta X$, and fall by $-2\Delta X$, in which case:

(a) $p_u + p_m + p_d = 1$.
(b) $E(\Delta X) = p_u 0 - p_m \Delta X - 2 p_d \Delta X = \hat{\mu}\Delta t$.
(c) $E(\Delta X^2) = p_u 0 + p_m \Delta X^2 + 4 p_d \Delta X^2 = \sigma^2 \Delta t + \hat{\mu}^2 (\Delta t)^2$.

The solution is then:

$$\begin{aligned} p_u &= \frac{7}{6} + \frac{M^2 + 3M}{2}, \\ p_m &= -\frac{1}{3} - M^2 - 2M, \\ p_d &= \frac{1}{6} + \frac{M^2 + M}{2}. \end{aligned}$$

If, instead, we have $p_d < 0$, then the price can either remain the same, rise by ΔX, and rise by $2\Delta X$. In this case:

(a) $p_u + p_m + p_d = 1$.
(b) $E(\Delta X) = p_u 2\Delta X + p_m \Delta X - p_d 0 = \hat{\mu}\Delta t$.
(c) $E(\Delta X^2) = 4 p_u \Delta X^2 + p_m \Delta X^2 + p_d 0 = \sigma^2 \Delta t + \hat{\mu}^2 (\Delta t)^2$.

The probabilities that solve this system are:

$$\begin{aligned} p_u &= \frac{1}{6} + \frac{M^2 - M}{2}, \\ p_m &= -\frac{1}{3} - M^2 + 2M, \\ p_d &= \frac{7}{6} + \frac{M^2 - 3M}{2}. \end{aligned}$$

Table 4.8 summarizes the above formulae.

Table 4.8 Formulae for the probabilities in the trinomial lattice

Case	p_u	p_m	p_d
Normal	$\frac{1}{6}+\frac{M^2+M}{2}$	$\frac{2}{3}-M^2$	$\frac{1}{6}+\frac{M^2-M}{2}$
High X	$\frac{7}{6}+\frac{M^2+3M}{2}$	$-\frac{1}{3}-M^2-2M$	$\frac{1}{6}+\frac{M^2+M}{2}$
Low X	$\frac{1}{6}+\frac{M^2-M}{2}$	$-\frac{1}{3}-M^2+2M$	$\frac{7}{6}+\frac{M^2-3M}{2}$

References

Clewlow L, Strickland C (1998) Implementing derivatives models. Wiley, Hoboken

Hull J, White A (1994) Numerical procedures for implementing term structure models I: single-factor models. J Deriv 2(1):7–16

Shreve SE (2005) Stochastic calculus for finance I: the binomial asset pricing model. Springer, Berlin

Trigeorgis L (1996) Real options. The MIT Press, Cambridge

Chapter 5
Finite Difference Methods

5.1 Introduction

These methods solve numerically the differential (pricing) equation of a derivative asset to value that derivative asset. In particular, the differential equation is transformed into a set of difference equations which are then solved iteratively.

Consider, for example, an American put option. For the sake of simplicity, assume that the underlying asset (worth S) is governed by the risk-neutral process:

$$dS = rSdt + \sigma SdZ; \tag{5.1}$$

assume further that the asset pays no dividend or convenience yield. The value of the option f must satisfy the following partial differential equation (PDE):

$$\frac{1}{2}\sigma^2 S^2 \frac{\partial^2 f}{\partial S^2} + rS\frac{\partial f}{\partial S} + \frac{\partial f}{\partial t} = rf. \tag{5.2}$$

Let 0 denote the current time and T the date of the option's maturity. We subdivide this time space into a finite number N of equally spaced time steps: $\Delta t = T/N$. We thus have $N + 1$ dates: 0, Δt, $2\Delta t$, $3\Delta t, \ldots, T$. We also subdivide the price space into a finite number M of equally spaced asset prices. Clearly, when $S = 0$ the put option reaches its highest value. Besides, there are a number of values of S which are so high that the put option becomes worthless; let S_{max} denote one such value. Thus, $\Delta S = S_{max}/M$. Consequently we have $M + 1$ asset prices: 0, ΔS, $2\Delta S$, $3\Delta S \ldots, S_{max}$; the current price corresponds to one of them.

Figure 5.1 illustrates this general approach. We set up a grid which comprises $(N + 1) \times (M + 1)$ points, or $N + 1$ dots on the horizontal (time) axis and $M + 1$ dots on the vertical (price) axis. The point (i, j) corresponds to time $i\Delta t$ and price $j\Delta S$. The value of the put option at that precise point is denoted by f_{ij}.

L.M. Abadie and J.M. Chamorro, *Investment in Energy Assets Under Uncertainty*, Lecture Notes in Energy 21, DOI: 10.1007/978-1-4471-5592-8_5,

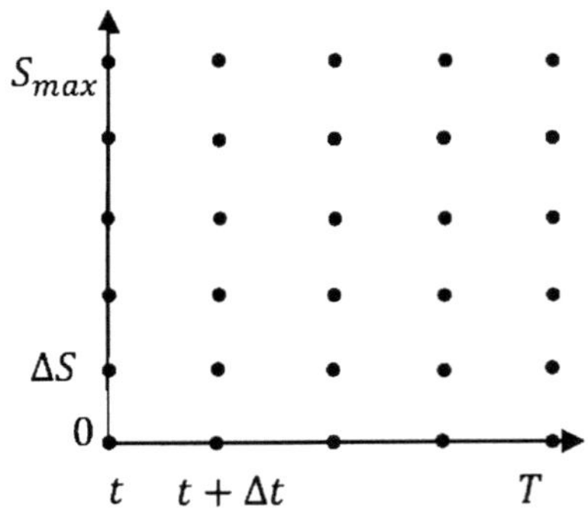

Fig. 5.1 Grid for finite difference methods

5.2 The Implicit Finite Difference Method

In an interior point (i, j), the partial derivative $\partial f/\partial S$ can be approximated by the ratio $\Delta f/\Delta S$ for which we have two possibilities:

$$\frac{\partial f}{\partial S} = \frac{f_{i,j+1} - f_{i,j}}{\Delta S} \quad or \quad \frac{\partial f}{\partial S} = \frac{f_{i,j} - f_{i,j-1}}{\Delta S}.$$

The first one is called the forward difference approximation, while the second one is known as the backward difference approximation. The implicit finite difference method averages both in a more symmetric approach:

$$\frac{\partial f}{\partial S} = \frac{f_{i,j+1} - f_{i,j-1}}{2\Delta S}. \tag{5.3}$$

Regarding the partial derivative $\partial f/\partial t$, we adopt a forward difference approximation. Thus the option value at time $i\Delta t$ is related to that at $(i+1)\Delta t$:

$$\frac{\partial f}{\partial t} = \frac{f_{i+1,j} - f_{i,j}}{\Delta t}. \tag{5.4}$$

As for the second-order partial derivative $\partial^2 f/\partial S^2$, the backward difference approximation of $\partial f/\partial S$ at point (i, j) was shown above: $\partial f/\partial S = (f_{i,j} - f_{i,j-1})/\Delta S$. The backward difference at $(i, j+1)$ is:

$$\frac{\partial^2 f}{\partial S^2} = \frac{\frac{f_{i,j+1}-f_{i,j}}{\Delta S} - \frac{f_{i,j}-f_{i,j-1}}{\Delta S}}{\Delta S} = \frac{f_{i,j+1} + f_{i,j-1} - 2f_{i,j}}{\Delta S^2}. \tag{5.5}$$

Substituting the above partial derivatives in the PDE and noting that $S = j\Delta S$ yields:

$$\frac{1}{2}\sigma^2 j^2 \Delta S^2 \frac{f_{i,j+1} + f_{i,j-1} - 2f_{i,j}}{\Delta S^2} + rj\Delta S \frac{f_{i,j+1} - f_{i,j-1}}{2\Delta S} + \frac{f_{i+1,j} - f_{i,j}}{\Delta t} = rf_{i,j} \tag{5.6}$$

for $i = 0, 1, 2, \ldots, N-1$ and $j = 1, 2, \ldots, M-1$. Rearranging we get:

$$a_j f_{i,j-1} + b_j f_{i,j} + c_j f_{i,j+1} = f_{i+1,j}, \tag{5.7}$$

where:

$$a_j \equiv \frac{1}{2} rj\Delta t - \frac{1}{2}\sigma^2 j^2 \Delta t, \quad b_j \equiv 1 + \sigma^2 j^2 \Delta t + r\Delta t, \quad c_j \equiv -\frac{1}{2} rj\Delta t - \frac{1}{2}\sigma^2 j^2 \Delta t.$$

Upon translating the PDE into a difference equation we use the boundary conditions of the option value. Since we deal with an American put option, its value at expiration (T) is $\max(K - S_T; 0)$. Therefore:

$$f_{N,j} = \max(K - j\Delta S; 0), \quad j = 0, 1, \ldots, M. \tag{5.8}$$

On the other hand, when the asset drops to zero the value of the option equals K:

$$f_{i,0} = K, \quad i = 0, 1, \ldots, N. \tag{5.9}$$

And the option value approaches zero as the asset price grows indefinitely:

$$f_{i,M} = 0, \quad i = 0, 1, \ldots, N. \tag{5.10}$$

The last three equations define the value of the put option along the three borders of the grid in Fig. 5.1, namely for $t = T$, $S = 0$, and $S = S_{max}$. We are naturally interested in the option value today, i.e. at time $t = 0$. We thus need to arrive at the value of f along the left border of the grid; to this end we use our difference equation Eq. (5.7) above:

$$a_j f_{i,j-1} + b_j f_{i,j} + c_j f_{i,j+1} = f_{i+1,j}.$$

We start from the points at the time prior to expiration $T - \Delta t$. With $i = N-1$ this equation provides $M-1$ simultaneous equations ($j = 1, 2, \ldots, M-1$):

$$a_j f_{N-1,j-1} + b_j f_{N-1,j} + c_j f_{N-1,j+1} = f_{N,j}. \tag{5.11}$$

The right-hand sides $f_{N,j}$ are known (right edge of the grid, $t = T$). In addition, from the lower and upper edges we derive, respectively:

$$f_{N-1,0} = K, \tag{5.12}$$

$$f_{N-1,M} = 0. \tag{5.13}$$

Therefore, the $M-1$ difference equations can be solved for the $M-1$ unknowns: $f_{N-1,1}, f_{N-1,2}, \ldots, f_{N-1,M-2}, f_{N-1,M-1}$.
Upon completing this task, each option value $f_{N-1,j}$ is compared with its corresponding intrinsic value $K - j\Delta S$ to check if it is optimal to exercise the option or rather to keep it alive. Thus, if $K - j\Delta S > f_{N-1,j}$ then early exercise at $T - \Delta t$ is optimal, and $f_{N-1,j} = K - j\Delta S$; otherwise the value $f_{N-1,j}$ is kept. The nodes of the grid for $t = T - 2\Delta t$ are treated in the same way, and so on and so forth. In the end

we reach the left edge of the grid and get $f_{0,1},\ldots,f_{0,M-1}$; one of them is the option price we sought after.

The implicit method always converges to the solution of the PDE as ΔS and Δt approach zero. An important drawback, however, is the requirement to solve $M-1$ difference equations to retrieve the $f_{i,j}$'s from the $f_{i+1,j}$'s. See Hull (1993).

5.3 The Explicit Finite Difference Method

The implicit method can be made simpler by assuming that the values of the partial derivatives $\partial f/\partial S$ and $\partial^2 f/\partial S^2$ at point (i, j) are the same as at $(i+1, j)$. These derivatives then become:

$$\frac{\partial f}{\partial S} = \frac{f_{i+1,j+1} - f_{i+1,j-1}}{2\Delta S}, \tag{5.14}$$

$$\frac{\partial^2 f}{\partial S^2} = \frac{f_{i+1,j+1} + f_{i+1,j-1} - 2f_{i+1,j}}{\Delta S^2}. \tag{5.15}$$

The difference equation changes accordingly:

$$f_{i,j} = a_j^* f_{i+1,j-1} + b_j^* f_{i+1,j} + c_j^* f_{i+1,j+1}, \tag{5.16}$$

where

$$a_j^* \equiv \frac{1}{1+r\Delta t}\left(-\frac{1}{2}rj\Delta t + \frac{1}{2}\sigma^2 j^2 \Delta t\right),$$
$$b_j^* \equiv \frac{1}{1+r\Delta t}\left(1 - \sigma^2 j^2 \Delta t\right),$$
$$c_j^* \equiv \frac{1}{1+r\Delta t}\left(\frac{1}{2}rj\Delta t + \frac{1}{2}\sigma^2 j^2 \Delta t\right).$$

This is the explicit finite difference method. Figure 5.2 displays the differences with the implicit method. The latter leads to a difference equation which relates three option values at time $i\Delta t$ to one value at a later time $(i+1)\Delta t$. Instead, the explicit method leads to another difference method which sets a relationship between an option value at $i\Delta t$ and three different ones at $(i+1)\Delta t$.

Finite difference methods are often used with $X \equiv \ln S$ as the underlying variable (instead of S). The PDE Eq. (5.2) then becomes:

$$\frac{1}{2}\sigma^2 \frac{\partial^2 f}{\partial X^2} + \left(r - \frac{1}{2}\sigma^2\right)\frac{\partial f}{\partial X} + \frac{\partial f}{\partial t} = rf. \tag{5.17}$$

The grid is then constructed with equally spaced values of ln S; this is slightly more efficient from a computational point of view. Besides, it has the advantage

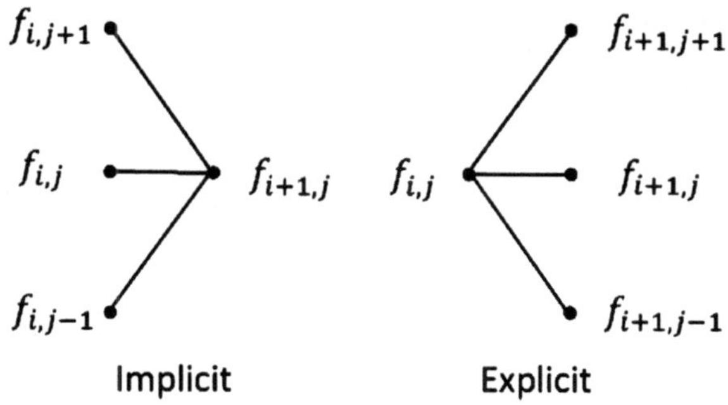

Fig. 5.2 Difference between the implicit method and the explicit method

that the coefficients $\{a_j, b_j, c_j\}$ and $\left\{a_j^*, b_j^*, c_j^*\right\}$ in the difference equations Eq. (5.7) and Eq. (5.16) are independent of j.

5.4 Relationship with Lattice Models

The explicit method resembles the lattice approach to a great extent. The terms in the expressions for $\left\{a_j^*, b_j^*, c_j^*\right\}$ can be interpreted as follows:

- $-\frac{1}{2}rj\Delta t + \frac{1}{2}\sigma^2 j^2 \Delta t$ The probability that, over the time interval Δt, the underlying asset price drops from $j\Delta S$ to $(j-1)\Delta S$
- $1 - \sigma^2 j^2 \Delta t$ The probability that, over the time interval Δt, the asset price remains unchanged at $j\Delta S$
- $\frac{1}{2}rj\Delta t + \frac{1}{2}\sigma^2 j^2 \Delta t$ The probability that the asset price rises from $j\Delta S$ to $(j+1)\Delta S$ over the time interval Δt.

Figure 5.3 illustrates this interpretation. The three probabilities sum to one. They give the expected change in the asset price over Δt as $rj\Delta S\Delta t = rS\Delta t$. This is the expected increment in a risk-neutral world. Provided Δt is small, they also give the variance of the asset price change over Δt as $\sigma^2 j^2 \Delta S \Delta t = \sigma^2 S^2 \Delta t$. This corresponds to the stochastic (GBM) process followed by the asset price S. Now, Eq. (5.15) evolves from $(i+1)\Delta t$ to $i\Delta t$ using a trinomial tree. $f_{i,j}$ is computed as the expected value at time $(i+1)\Delta t$ under risk neutrality discounted at the riskless interest rate, i.e. $1/(1+r\Delta t)$.

The explicit method works well provided the three "probabilities" above are positive. This is not always the case; for example, high values of the asset price S can give rise to negative option values or other inconsistencies. In other words, the explicit method not necessarily converges to the solution of the PDE. In many cases, it suffices to develop the lattice for ln S (instead of S) to overcome this problem.

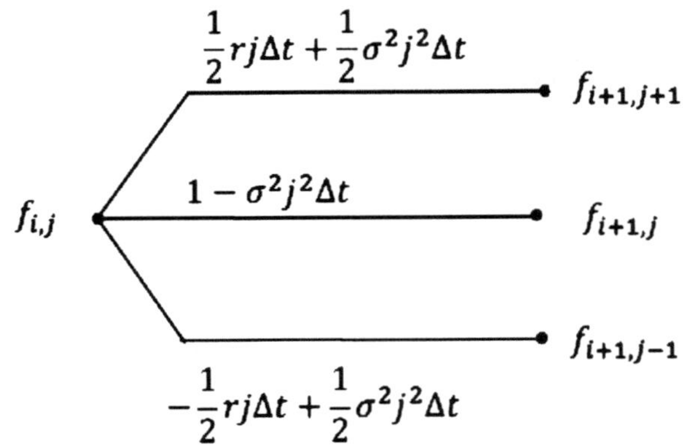

Fig. 5.3 Interpretation of the explicit method as a trinomial lattice

Finite difference methods can be used for pricing the same derivative assets as binomial lattices. They can deal with both European and American options; nonetheless, they are not so well suited for valuing path-dependent options. They can also handle several sources of risk (by building a multidimensional grid), though this entails a sizeable increase in computing time.

5.5 Example 1: Valuation of a European Real Option

The standard Black-Scholes PDE for the value of a European option $f(S,t)$ is:

$$\frac{1}{2}\sigma^2 S^2 f_{SS} + rSf_S + f_t - rf = 0. \tag{5.18}$$

An opportunity to invest is akin to a call option, so the boundary conditions that apply in this case are:

- $t = T{:}f(S,T) = max(V(S_T) - I;0)$ (5.19)
- $S = 0{:}f(0,t) = 0$ (5.20)

Though the implicit method is more robust, here we apply the explicit method for simplicity. Our two state variables, S and t, are discretized in N and M subintervals, respectively. Substituting the discrete approximations for the partial derivatives we get the difference equation Eq. (5.16):

$$f_{i,j} = a_j^* f_{i+1,j-1} + b_j^* f_{i+1,j} + c_j^* f_{i+1,j+1}.$$

Thus, we can derive the time-i values from the (known) values at time $i+1$. Since we have a boundary condition which provides us with starting values for $i = N$, we can proceed backward to $i = 0$.

Table 5.1 shows the solution to the option valuation problem under the assumptions $T = 1$, $r = 0.10$, $I = 10$, and $\sigma = 0.20$; see Cortazar (2001). The

Table 5.1 Value of a European real option using the explicit method

j	S	1.0	0.9	0.8	0.7	0.6	0.5	0.4	0.3	0.2	0.1	0	a_j^*	b_j^*	c_j^*
20	2.0	10	10	10	10	10	10	10	10	10	10	10	0.6931	−0.594	0.8911
19	1.9	9.2	9.3	9.2	9.3	9.2	9.2	9.2	9.2	9.1	9.1	9	0.6208	−0.439	0.8089
18	1.8	8.6	8.4	8.5	8.4	8.4	8.3	8.3	8.2	8.2	8.1	8	0.5525	−0.293	0.7307
17	1.7	7.6	7.7	7.6	7.6	7.5	7.4	7.4	7.3	7.2	7.1	7	0.4881	−0.154	0.6564
16	1.6	6.8	6.7	6.7	6.6	6.6	6.5	6.4	6.3	6.2	6.1	6	0.4277	−0.023	0.5861
15	1.5	5.9	5.8	5.7	5.7	5.6	5.5	5.4	5.3	5.2	5.1	5	0.3713	0.0990	0.5198
14	1.4	4.9	4.8	4.8	4.7	4.6	4.5	4.4	4.3	4.2	4.1	4	0.3188	0.2139	0.4574
13	1.3	4.0	3.9	3.8	3.7	3.6	3.5	3.4	3.3	3.2	3.1	3	0.2703	0.3208	0.3990
12	1.2	3.01	2.9	2.8	2.7	2.6	2.5	2.4	2.3	2.2	2.1	2	0.2257	0.4198	0.3446
11	1.1	2.11	2.0	1.9	1.8	1.7	1.6	1.5	1.3	1.2	1.1	1	0.1851	0.5109	0.2941
10	1.0	1.30	1.2	1.1	1.0	0.9	0.8	0.7	0.6	0.4	0.2	0	0.1485	0.5941	0.2475
9	0.9	0.67	0.6	0.5	0.4	0.4	0.3	0.2	0.1	0.1	0.0	0	0.1158	0.6693	0.2050
8	0.8	0.26	0.2	0.2	0.1	0.1	0.1	0.0	0.0	0.0	0.0	0	0.0871	0.7366	0.1663
7	0.7	0.07	0.0	0.0	0.0	0.0	0.0	0.0	0.0	0.0	0.0	0	0.0624	0.7960	0.1317
6	0.6	0.01	0.0	0.0	0.0	0.0	0.0	0.0	0.0	0.0	0.0	0	0.0416	0.8475	0.1010
5	0.5	0.00	0.0	0.0	0.0	0.0	0.0	0.0	0.0	0.0	0.0	0	0.0248	0.8911	0.0743
4	0.4	0.00	0.0	0.0	0.0	0.0	0.0	0.0	0.0	0.0	0.0	0	0.0119	0.9267	0.0515
3	0.3	0.00	0.0	0.0	0.0	0.0	0.0	0.0	0.0	0.0	0.0	0	0.0030	0.9545	0.0327
2	0.2	0.00	0.0	0.0	0.0	0.0	0.0	0.0	0.0	0.0	0.0	0	−0.002	0.9743	0.0178
1	0.1	0.00	0.0	0.0	0.0	0.0	0.0	0.0	0.0	0.0	0.0	0	−0.003	0.9861	0.0069
0	0	0	0	0	0	0	0	0	0	0	0	0	0.0000	0.9901	0.0000

Table 5.2 Solution from finite differences and analytical solution ($r = 0.10$, $\sigma = 0.20$)

	$S=0$	$S=0.4$	$S=0.8$	$S=1.2$	$S=1.6$	$S=2.0$
Analytic	0.0000	0.0000	0.0279	0.3026	0.6953	1.0952
Finite D.	0.0000	0.0000	0.2555	0.3012	0.6828	1.000

coefficients a_j^*, b_j^*, and c_j^* appear at the right; they are needed for the computations in the left part. The contiguous column displays the value of the real option along the (time) boundary:

$$f(S,T) = max(V(S_T) - I; 0).$$

The former difference equations allow compute one column after another until reaching $t = 0$. This column shows the value of the option for each initial (asset) price when the option's maturity is one year. Note that the heading of these columns shows the time to maturity (from right to left, we read 0, 0.1, …, 0.9, 1.0).

Instead, Table 5.2 displays the exact option values alongside those approximated by the explicit method for comparison.

5.6 The Crank-Nicolson Method

Both the explicit and the implicit method involve considering three nodes at one date and one node at another date (with the two dates separated by Δt). And both have important advantages and shortcomings. It is possible to devise a better scheme by considering the three points at the two dates.

The approximation to the second-order derivative in the implicit method is:

$$\frac{f_{i,j+1} - 2f_{i,j} + f_{i,j-1}}{\Delta S^2}.$$

The three points are dated at $i\Delta t$. The approximation to the same derivative in the explicit method is:

$$\frac{f_{i+1,j+1} - 2f_{i+1,j} + f_{i+1,j-1}}{\Delta S^2}.$$

The three points are dated at $(i+1)\Delta t$. We can construct a linear convex combination of the two approximations:

$$\frac{1}{\Delta S^2}\left[\lambda\left(f_{i,j+1} - 2f_{i,j} + f_{i,j-1}\right) + (1-\lambda)\left(f_{i+1,j+1} - 2f_{i+1,j} + f_{i+1,j-1}\right)\right], \tag{5.21}$$

for $0 \le \lambda \le 1$. The implicit method corresponds to $\lambda = 1$, whereas for $\lambda = 0$ we get the explicit method. For $\lambda = 1/2$ we get the Crank-Nicolson method.

The change in the discrete approximation to the second-order partial derivative affects the difference equation describing the scheme. To begin with, the sum of the six products equals the sum of the two elements:

$$f_{i+1,j} + f_{i,j} = a_j f_{i,j-1} + b_j f_{i,j} + c_j f_{i,j+1} + a_j^* f_{i+1,j-1} + b_j^* f_{i+1,j} + c_j^* f_{i+1,j+1}. \quad (5.22)$$

Now let us define:

$$g_{i+1,j} \equiv f_{i+1,j} - a_j^* f_{i+1,j-1} - b_j^* f_{i+1,j} - c_j^* f_{i+1,j+1}. \quad (5.23)$$

Then we have the following difference equation:

$$g_{i+1,j} = a_j f_{i,j-1} + b_j f_{i,j} + c_j f_{i,j+1} - f_{i,j}. \quad (5.24)$$

The main advantage of this method rests on the lower truncation errors relative the explicit and implicit schemes. Therefore, a numerical solution with suitable accuracy can be achieved at less computational effort. See Brandimarte (2002) and Wilmott (1998).

5.7 Example 2: Valuation of an American Put Option

Assume the parameter values: $S_0 = 10$, $\alpha - \lambda = 0.05$, $\sigma = 0.40$, $T = 10$, and $r = 0.10$. The put option will be in-the-money when the exercise price (I) satisfies the condition $I > S_t$; the option payoff will be:

$$W_{ij} = max\left(I - S_{ij}; 0\right).$$

We assume $I = 40$. The option value has been computed (by the Crank-Nicolson method) using $\Delta S = 1$, $\Delta t = 1/100$, and $S_{max} = 400$. Table 5.3 displays the results. For lower values of S, the value of the put option is higher. Doubling the time to maturity enhances the option value. The effect of doubling volatility is even stronger.

Table 5.3 Value of an American put option by finite differences

S_0	σ	T (years)	Crank-Nicolson
38	0.20	1	3.2497
38	0.20	2	3.7463
38	0.40	1	6.1508
38	0.40	2	7.6725
40	0.20	1	2.3122
40	0.20	2	2.8849
40	0.40	1	5.3146
40	0.40	2	6.9211
42	0.20	1	1.6147
42	0.20	2	2.2121
42	0.40	1	4.5847
42	0.40	2	6.2480

Table 5.4 Valuation of a long-lived American put by Crank-Nicolson method

Volatility	Maturity (years)	Crank-Nicolson
0.20	10	3.9677
0.30	10	7.3139
0.40	10	10.7262
0.20	25	4.1881
0.30	25	7.9844
0.40	25	11.8466
0.20	50	4.2169
0.30	50	8.1129
0.40	50	12.0475

5.8 Example 3: Valuation of a Long-Term American Put Option

This case is similar to Example 2 above. Now however, the option's maturity T is alternatively 10, 20 or 50 years. We adopt $S_0 = I = 40$. The results assume $\Delta S = 1$, $S_{max} = 800$, and $\Delta t = 1/50$. Table 5.4 shows the results. The qualitative results remain the same as in Table 5.3.

Chapter 6 includes several comparisons of the results from the three numerical methods for different options.

References

Brandimarte P (2002) Numerical methods in finance. Wiley, New York

Cortazar G (2001) Simulation and numerical methods in real options valuation. In: Schwartz ES, Trigeorgis L (eds) Real options and investment under uncertainty. The MIT Press, London

Hull J (1993) Options, futures, and other derivative securities, 2nd edn. Prentice Hall, Englewood Cliffs

Wilmott P (1998) Derivatives. The theory and practice of financial engineering. Wiley, Chichester

Chapter 6
Monte Carlo Simulation

6.1 Introduction

We can always value a derivative asset as if the world were risk neutral as long as the expected rate of return on the underlying asset is consistent with that assumption (for example, with trading on the futures markets). At one level, we regularly address the valuation of options on stocks, indexes, currencies, or commodities under deterministic riskless interest rates. Nonetheless, the risk-free rate can itself be stochastic, in which case it must be treated accordingly, i.e. the same way that any other underlying variable (Wilmott 1998). Below we consider an example where the interest rate is stochastic.

Monte Carlo simulation is a general and powerful technique. One major shortcoming is the relative inefficiency (in comparison to other numerical techniques) when there is a low number of underlying variables (say, three or four). In addition, its application to American options is far from trivial because of the early exercise possibility. The optimal time to exercise the option can only be identified after checking all the points in the (S, t) space up to the option's expiration. In particular, the option holder must weight the intrinsic value of exercising immediately against the alternative of waiting for one more interval Δt thus keeping the option alive.

6.2 The Basic Setup: Only One GBM Underlying Variable

Consider a European-type contingent claim on an underlying asset worth S; assume that the claim conveys a payoff f_T at time T. To value this claim we must set the expected growth rate of S at the appropriate level, i.e. that in a risk-neutral world (r if it is a traded security with no dividends), while keeping the volatility equal to σ. Thus, the risk-neutral process that governs S is:

$$dS = rSdt + \sigma SdZ. \tag{6.1}$$

L.M. Abadie and J.M. Chamorro, *Investment in Energy Assets Under Uncertainty*, Lecture Notes in Energy 21, DOI: 10.1007/978-1-4471-5592-8_6,

The value of the claim is just the expected payoff discounted to the present (time t) at the riskless rate. Analytically, the value f of an asset that pays f_T at time T is given by:

$$f = e^{-r(T-t)}E^Q(f_T), \tag{6.2}$$

where E^Q stands for the mathematical expectation operator under risk neutrality (not in the physical world).

Monte Carlo simulation allows computing the value of a European option from this equation. We can assess this value following a few simple steps:

(a) Simulate the risk-neutral process (as shown below) starting from the current price of the underlying asset, S_0, over the whole time horizon until expiration of the option. This provides a realization of the path of the asset price.
(b) Compute the option's terminal payoff f_T under this realization (a simulation run).
(c) Develop many more (say, by the thousands) similar realizations over the time horizon and assess their respective terminal values.
(d) Compute the average option's payoff from all the realizations, $E^Q(f_T)$.
(e) Take the discounted value of that average; this is the option value f.

A prior ingredient to the first part is generation of random numbers from a standard normal distribution $N(0, 1)$ (or some suitable approximation). Then we must revise or update the asset price at each time step Δt using these random increments. This is thus a discrete-time method.

A possible choice is adopting:

$$\Delta S = rS\Delta t + \sigma S\sqrt{\Delta t}\varepsilon, \tag{6.3}$$

where ΔS denotes the change in S over Δt, and ε stands for a random sample from a $N(0, 1)$ distribution. This procedure to simulate the time path of S is called Euler's method. We merely substitute the latest known value of S in the right hand of the equation to derive ΔS and hence the next value of S. Successive increments ΔS provide a simulated path of S and enable to compute a sample terminal payoff of the derivative asset f_T.

Depending on the size of Δt we incur a discretization error. The type of discretization in Eq. (6.3) can be easily applied to any stochastic differential equation. Yet as a major drawback it has a discretization error of order $O(\Delta t)$. If, in order to reduce the latter, we take Δt very small, we will incur a higher computation cost. Fortunately there are better approximations, e.g. Milstein's method, which has an error of order $O(\Delta t^2)$.

In the particular case of a GBM, however, it is possible to find a discretization algorithm which is both exact and simple. The risk-neutral GBM followed by $X_t \equiv \ln S_t$ is:

$$d(\ln S_t) = \left(r - \frac{\sigma^2}{2}\right)dt + \sigma dZ_t. \tag{6.4}$$

We can integrate this differential equation exactly; the result is:

$$S_t = S_0 e^{\left(r-\frac{\sigma^2}{2}\right)t+\sigma\int_0^t dZ}. \tag{6.5}$$

Now, over a time step Δt we have:

$$S_{t+\Delta t} = S_t + \Delta S = S_t e^{\left(r-\frac{\sigma^2}{2}\right)\Delta t+\sigma\sqrt{\Delta t}\varepsilon}. \tag{6.6}$$

Note that this is an exact expression. Therefore, Δt need not be small. Indeed, if there is just one option's payoff which depends only on the terminal value of the asset then we can simulate the latter in a great leap using a time step of length T. Note, however, that there remains the error that can arise from using a finite number of random numbers.

6.2.1 Use of Random Numbers

In the above case of a simple European option on an asset following a GBM we can leave aside any concern about the suitable Δt since we have an exact formula. We must remain concerned, however, about simulating a high enough number of random paths so as to derive an accurate value of the option. After all, we only simulate a finite number of infinite potential paths. Exploring a number N of realizations of the asset price involves an error $O(1/\sqrt{N})$.

For more complex derivative assets (like path-dependent options), the choice of the time step Δt is not trivial. We can make errors $O(\Delta t)$ when approximating continuous-time processes in discrete time. An estimate of the derivative price thus requires a total number of computations $O(N/\Delta t)$; this serves also as a gauge of the time consumed in estimating that price. So the error in the price of the derivative asset is $O(max(1/\sqrt{N}, \Delta t))$, i.e. the worst of both error sources. In order to minimize it while fixing the time of computation at $O(N/\Delta t) = K$ we must choose $N = O(K^{2/3})$ and $\Delta t = O(K^{-1/3})$; see Jäckel (2002).

6.2.2 Example 1: Comparison with a GBM Annuity

We know that a 20-year annuity over a GBM-behaved income S has a present value:

$$V(S) = \frac{S}{\alpha - \lambda - r}\left[e^{(\alpha-\lambda-r)\tau_2} - e^{(\alpha-\lambda-r)\tau_1}\right]. \tag{6.7}$$

For $S = 10$, $\alpha - \lambda = 0.05$, $r = 0.10$, $\tau_1 = 0$, and $\tau_2 = 20$ we compute $V(10) = 126.4241$. Now we are going to check this result (which does not depend on σ) through Monte Carlo simulation.

We will use the following discretization:

$$S_{t+\Delta t} = S_t e^{\left(\alpha-\lambda-\frac{1}{2}\sigma^2\right)\Delta t+\sigma\sqrt{\Delta t}\varepsilon_t}, \tag{6.8}$$

where ε_t is a standardized Gaussian white noise. In this case, owing to the properties of the log-normal distribution, the expected value is $E_t^Q(S_{t+\Delta t}) = S_t e^{(\alpha-\lambda)\Delta t}$. Nonetheless, if the time step Δt is small, we could also use the following specification (with a small loss in precision):

$$S_{t+\Delta t} = S_t + (\alpha - \lambda)S_t\Delta t + \sigma S_t\sqrt{\Delta t}\varepsilon_t. \tag{6.9}$$

If we run m simulations and, in each of them, we have subdivided the time interval into n steps ($T = n\Delta t$) then we will have $m \times n$ values S_{ij}, with $i = 1, 2, \ldots, m$ and $j = 1, 2, \ldots, n$. Each simulation provides a value:

$$V_i = \sum_{j=1}^{j=n} S_{ij}\Delta t e^{-rj\Delta t} = \frac{T}{n}\sum_{j=1}^{j=n} S_{ij}e^{-rj\Delta t}. \tag{6.10}$$

Since the asset price S_t refers to the value of one unit over a year, we multiply it by Δt in the above Eq. (6.10). To estimate the value of the annuity by simulation we average across all the values:

$$V = \frac{1}{m}\sum_{i=1}^{i=m} V_i. \tag{6.11}$$

We know that the expected value after 20 years ($S_0 = 10$), must be:

$$E_0^Q(S_{20}) = S_0 e^{(\alpha-\lambda)20} = 10e = 27.8128.$$

On the other hand, the variance must be ($\sigma = 0.40$):

$$Var(S_t) = S_0^2 e^{2(\alpha-\lambda)t}\left(e^{\sigma^2 t} - 1\right) \to Var(S_{20}) = 10^2 e^2\left(e^{3.2} - 1\right) = 17388.32.$$

Thus the standard deviation equals $Std(S_t) = 131.8648$.

The value of σ should not affect the value of the annuity, yet it will affect the accuracy of our estimates. Table 6.1 shows the sensitivity to changes in σ under the assumptions $m = 20{,}000$ and $n = 2{,}000$ (which implies $\Delta t = 1/100$). The differences between the analytical solutions and those derived by simulation are very small.

Table 6.1 Sensitivity of the value of the annuity to changes in volatility

σ	Estimated solution			Analytic solution		
	V	S_{20}	$Std(S_t)$	V	S_{20}	$Std(S_t)$
0.40	126.9782	27.4440	135.1511	126.4241	27.1828	131.8648
0.30	126.6148	27.1067	63.9674	126.4241	27.1828	61.0836
0.20	126.4886	27.0957	30.4325	126.4241	27.1828	30.0925
0.10	126.4488	27.1718	12.7008	126.4241	27.1828	12.7905

6.2.3 *Example 2: A GBM Annuity with Jump (Convergence to Perpetual Annuity)*

Assume $S_0 = 10$, $\alpha - \lambda = 0.05$, $\sigma = 0.40$, $r = 0.10$, $T = 60$, $m = 20{,}000$, and $n = 6{,}000$. The mean arrival rate of the jump process is $\eta = 0.05$, and the percentage fall in the value of S upon event occurrence is $\phi = 1$ (obviously $0 \leq \phi \leq 1$). In the case of an annuity between τ_1 and τ_2 with arrival rate η the analytic solutionfor its value is:

$$V(S_t) = \frac{S_t}{\alpha - \lambda - r - \eta}\left[e^{(\alpha-\lambda-r-\eta)\tau_2} - e^{(\alpha-\lambda-r-\eta)\tau_1}\right]. \tag{6.12}$$

In particular, with $\tau_1 = 0$, $\tau_2 = 20$ and $\eta = 0.05$ we compute $V(10) = 86.4665$.

Now we check this formula by means of Monte Carlo simulation with 100 steps per year and 20,000 paths. In this case the probability that S_t drops to zero at any time step is $\eta\Delta t = 0.0005$. We get $V(10) = 86.4952$, which is very close to the analytic solution. In particular, for each time step we have:

- With probability $1 - \eta\Delta t = 0.9995$: $S_{t+\Delta t} = S_t e^{\left(\alpha-\lambda-\frac{1}{2}\sigma^2\right)\Delta t+\sigma\sqrt{\Delta t}\varepsilon_t}$,
- With probability $\eta\Delta t = 0.0005$: $S_{t+\Delta t} = 0$.

The value of a perpetual annuity with $\tau_1 = 0$ and $\tau_2 \to \infty$ will be

$$V(S_0) = \frac{S_0}{(r+\eta) - (\alpha+\lambda)} = 10S_0.$$

Note that the above formula is meaningful provided $(r+\eta) > (\alpha+\lambda)$. For $S_0 = 10$ we get $V(10) = 100$. Running Monte Carlo simulations with 20,000 paths and 100 steps per year we get the results displayed in Table 6.2. The value of the annuity grows with the time to maturity.

Table 6.2 Simulated values of the perpetual annuity

Years	20	30	40	50
$V(10)$	86.4952	96.3981	97.0835	99.5299

6.2.4 Example 3: A GBM Annuity with Jump ($\phi = 0.50$)

As before, $S_0 = 10$, $\alpha - \lambda = 0.05$, $\sigma = 0.40$, $r = 0.10$, $T = 60$, $m = 20{,}000$, $n = 6{,}000$, and $\eta = 0.05$. Now however, $\phi = 0.50$. In this case, the present value of an annuity between τ_1 and τ_2 is:

$$V(S_t) = \int_{\tau_1}^{\tau_2} E_0^Q(S_t)e^{-rt}dt = \frac{S_t}{\alpha - \lambda - r - \phi\eta}\left[e^{(\alpha-\lambda-r-\phi\eta)\tau_2} - e^{(\alpha-\lambda-r-\phi\eta)\tau_1}\right]. \tag{6.13}$$

In particular, with $\tau_1 = 0$ and $\tau_2 = 20$ we get $V(10) = 10.3583S_t = 103.5826$.

Through simulation, with 100 steps per year and 40,000 paths we compute $V(10) = 103.4420$; this is very close to the analytic solution. Now we have used:

- With probability $1 - \eta\Delta t = 0.9995$: $S_{t+\Delta t} = S_t e^{\left(\alpha-\lambda-\frac{1}{2}\sigma^2\right)\Delta t+\sigma\sqrt{\Delta t}\varepsilon_t}$,
- With probability $\eta\Delta t = 0.0005$: $S_{t+\Delta t} = (1 - \phi)S_t$.

At each step we have taken random samples from a uniform distribution to decide if there is a jump or not. Thus, if the sample value is lower than or equal to $\eta\Delta t = 0.0005$ then a jump takes place and $S_{t+\Delta t} = (1 - \phi)S_t$. Otherwise, if the sample value is higher than $\eta\Delta t = 0.0005$ then we take a random sample from a Normal distribution (ε_t) and compute $S_{t+\Delta t} = S_t e^{\left(\alpha-\lambda-\frac{1}{2}\sigma^2\right)\Delta t+\sigma\sqrt{\Delta t}\varepsilon_t}$.

6.2.5 Example 4: Valuation of a European Option by Simulation

We know that, when the underlying asset earns a convenience yield $\delta = r - (\alpha - \lambda)$, the value of a call option (C) and a put option (P) with exercise price K and maturity T are given by:

$$C = S_0 e^{((\alpha-\lambda)-r)T}N(d_1) - Ke^{-rT}N(d_2), \tag{6.14}$$

$$P = Ke^{-rT}N(-d_2) - S_0 e^{((\alpha-\lambda)-r)T}N(-d_1), \tag{6.15}$$

where

$$d_1 = \frac{\ln\frac{S_0}{K} + \left[(\alpha - \lambda) + \frac{1}{2}\sigma^2\right]T}{\sigma\sqrt{T}}, d_2 = 1 - d_1 = \frac{\ln\frac{S_0}{K} + \left[(\alpha - \lambda) - \frac{1}{2}\sigma^2\right]T}{\sigma\sqrt{T}}.$$

In our case we assume $S_0 = 10$, $\alpha - \lambda = 0.05$, $\sigma = 0.40$, $r = 0.10$, $T = 5$, and $K = 10$. The analytic solutions are $C = 3.3392$ and $P = 1.6165$, respectively.

To check these numerical results we run $m = 1{,}000{,}000$ simulations using:

$$S_i = S_0 e^{\left(\alpha-\lambda-\frac{1}{2}\sigma^2\right)\Delta t+\sigma\sqrt{\Delta t}\varepsilon_i}.$$

Given that the final distribution is known, it is not necessary to subdivide the five-year period into pre-determined steps. Indeed, this way we avoid possible errors due to an unsuitable discretization when one uses a step Δt which is not short enough. For the computation of C and P we use:

$$C = \frac{1}{m}\sum_{i=1}^{i=m} \max(S_i - K; 0)e^{-rT}, \tag{6.16}$$

$$P = \frac{1}{m}\sum_{i=1}^{i=m} \max(K - S_i; 0)e^{-rT}. \tag{6.17}$$

The values from Monte Carlo simulation are $C = 3.3396$ and $P = 1.6147$, again very close to those derived from Eqs. (6.14) and (6.15).

6.2.6 Variance Reduction Techniques

Let k denote the standard deviation of the option value using a single simulation. After N simulations the standard deviation of the error is $k/\sqrt{N}$. Thus, we need to run 100 times as many simulations to improve accuracy by a factor of 10. In principle, it would be necessary to get a huge number of simulation runs to compute an accurate estimate of f. Not surprisingly, there have been several attempts for making convergence faster. Improved sampling methods are part of them. See Hull (1993) and Brandimarte (2002).

6.2.7 Example 5: Valuation of a European Option by Simulation with Sobol Low-discrepancy Sequences

Again we run $m = 1{,}000{,}000$ simulations using:

$$S_i = S_0 e^{\left(\alpha-\lambda-\frac{1}{2}\sigma^2\right)\Delta t+\sigma\sqrt{\Delta t}\varepsilon_i}$$

as in Example 4. Now, however, we start from Sobol low-discrepancy sequences (Brandimarte 2002); as before, we derive the shock terms ε_i using the inverse of the Normal distribution. The option values are $C = 3.3392$ and $P = 1.6165$. They coincide with those from the analytic solution with accuracy up to the fourth decimal figure.

6.3 Monte Carlo Simulation and American Options Valuation

There have been several efforts to extend Monte Carlo simulation techniques for solving American-type options. Longstaff and Schwartz (2001) present a simple, yet powerful new approach for valuing American options, the so-called Least Squares Monte Carlo (LSM) approach. At any time prior to the option maturity, the holder compares the payoff from immediate exercise with the expected payoff from continuation (i.e., from keeping the option unexercised). Maximising the value of the option entails choosing the optimal time to exercise it, which is the first time that the exercise value surpasses the continuation value. The optimal exercise strategy is thus determined by the conditional expected payoff from keeping the option alive. In order to estimate this conditional expectation, first they run a number of simulation paths of the state variables; second, they determine the optimal exercise time by backward induction. At any point in time (starting from the end), each path generates one observation on the optimality of exercising or not for that path. Using cross-sectional regressions it is possible to estimate when it is optimal to exercise for given date and state variable values by using least squares, and solve recursively backward. Specifically, they regress the ex-post realized payoffs from continuation on functions of the values of the state variables. The fitted value from this regression provides a direct estimate of the conditional expectation function. By estimating this function for each exercise date, we obtain a complete specification of the optimal exercise strategy along each path. With this specification, American options can then be valued accurately by simulation.

Much research has focused on the development of methods to compute approximations to the optimal exercise policy. These methods often incur unknown approximation errors and are limited by a lack of error bounds (see Broadie and Detemple 2004).

6.3.1 Example 6: Valuation of an American Option by Simulation

This case corresponds to Example 3 of Chap. 4 (Sect. 4.2.6) for $T = 10$, $S_0 = 10$, $\alpha - \lambda = 0.05$, $\sigma = 0.40$, $r = 0.10$, $I = 280$, $\Delta t = 1/100$, $\tau_1 = 0$ and $\tau_2 = 20$. The investment option is in the money when $12.64\, S_t < I$; this element must be included in the program designed for computing the value of the option by the LSMC method. Instead of the value 33.0342 that we derived from the binomial lattice (see Table 4.1), now we get 33.1921 by LSMC.

To get this figure we have generated a number m = 1,000,000 of paths for S_t, each consisting of 100 time steps per year (i.e. each comprising n = 1,000 steps). For each one of the 100 million values of S_{ij} we compute the corresponding investment value:

$$V(S_{ij}) = \frac{S_{ij}}{\alpha - \lambda - r}\left[e^{(\alpha-\lambda-r)\tau_2} - 1\right] = 12.64 S_{ij}. \tag{6.18}$$

Then we get the option payoff:

$$W_{ij} = max\left(V\left(S_{ij}\right) - I; 0\right).$$

As basic functions (Longstaff and Schwartz 2001; Brandimarte 2002), we have used the components of a second-degree polynomial.

6.3.2 Example 7: Valuation of an American Option by Simulation (Decreasing Investment Cost)

This case corresponds to Example 6 in Chap. 4 (Sect. 4.2.9) for $T = 10$. As in the last case, $S_0 = 10$, $\alpha - \lambda = 0.05$, $\sigma = 0.40$, $r = 0.10$, $I = 280$, $\Delta t = 1/100$, $\tau_1 = 0$ and $\tau_2 = 20$. Now, though:

$$I_t = I_0 e^{\beta t} = 280 e^{-0.025t} = 280 e^{-0.025 j\Delta t};$$
$$W_{ij} = max\left(V\left(S_{ij}\right) - I_0 e^{-0.025 j\Delta t}; 0\right).$$

Following the LSMC approach we get a value 37.9276. If, instead, we followed the lattice approach as in Sect. 4.2.9, we would come up with a value 38.0593. In sum, the results from the two approaches for valuing American options are pretty similar.

6.3.3 Example 8: The American Put Option by LSMC, Binomial Lattice, and Finite Differences

In this example we compare the results from three different numerical methods. Here the option will be in-the-money when the exercise price (I) satisfies the condition $I > S_t$; the option payoff will be:

$$W_{ij} = max\left(I - S_{ij}; 0\right).$$

We assume $I = 40$ and $r = 0.06$. The value derived from finite differences (Crank-Nicolson) has been computed using $\Delta S = 1$, $\Delta t = 1/100$, and $S_{max} = 400$. Table 6.3 displays the results. For the cases in the table, the binomial lattice tends to give relatively higher values; nonetheless, the difference with respect to the lowest value is typically close to 0.5 %.

Table 6.3 Valuation of an American put following three different approaches

S_0	σ	T (years)	LSMC	Binomial	Crank-Nicol.
38	0.20	1	3.2469	3.2603	3.2497
38	0.20	2	3.7327	3.7521	3.7463
38	0.40	1	6.1480	6.1649	6.1508
38	0.40	2	7.6551	7.6804	7.6725
40	0.20	1	2.3063	2.3165	2.3122
40	0.20	2	2.8738	2.8877	2.8849
40	0.40	1	5.3068	5.3105	5.3146
40	0.40	2	6.8971	6.9186	6.9211
42	0.20	1	1.6097	1.6223	1.6147
42	0.20	2	2.1987	2.2168	2.2121
42	0.40	1	4.5747	4.6017	4.5847
42	0.40	2	6.2276	6.2578	6.2480

6.3.4 Example 9: Long-Term American Put (Three Approaches)

We consider a case similar to Example 8 above. Now, however, the option's maturity is alternatively 10, 25 or 50 years. We adopt $S_0 = I = 40$, and $r = 0.06$. The results from LSMC rest on 50 steps per year and 100,000 paths. Those from binomial lattices and finite differences (Crank-Nicolson), instead, assume $\Delta t = 1/100$. The latter further assumes $\Delta S = 1$, $S_{max} = 800$, and $\Delta t = 1/50$. Table 6.4 shows the results.

6.3.5 Example 10: An IGBM Underlying Variable

This case corresponds to Example 9 in Chap. 4 (Sect. 4.3.3). Again, $S_0 = 110$, $\frac{kS_m}{k+\lambda} = 90$, $k + \lambda = 0.30$, $T = 10$, $\Delta t = 1/100$, while $I = 1{,}500$, $\tau_1 = 0$, $\tau_2 = 20$,

Table 6.4 Valuation of an American put following three different approaches

Volatility	Maturity	LSMC	Binomial	Crank-Nicolson
0.20	10	3.9377	3.9677	3.9677
0.30	10	7.2679	7.3103	7.3139
0.40	10	10.6880	10.7212	10.7262
0.20	25	4.1722	4.1857	4.1881
0.30	25	7.9810	7.9806	7.9844
0.40	25	11.8558	11.8480	11.8466
0.20	50	4.1976	4.2140	4.2169
0.30	50	8.1051	8.1110	8.1129
0.40	50	12.0981	12.0776	12.0475

$r = 0.02$, and $\sigma = 0.25$. First we generate 100,000 simulation paths; each one consists of 1,000 steps:

$$S_{t+\Delta t} = S_m\left(1 - e^{-k\Delta t}\right) + S_t e^{-k\Delta t} + \sigma S_t \sqrt{\Delta t}\varepsilon_t. \tag{6.19}$$

We get an option value 70.4631. It is close to 70.6181 that we computed before (see Table 4.6). In this case, the level of costs in each node is given by:

$$V\left(S_{ij}\right) = \frac{S_{ij} - \frac{kS_m}{\lambda+k}}{\lambda + k + r}\left[e^{-(\lambda+k+r)\tau_1} - e^{-(\lambda+k+r)\tau_2}\right] + \frac{kS_m}{r(\lambda+k)}\left[e^{-r\tau_1} - e^{-r\tau_2}\right]. \tag{6.20}$$

The corresponding cash flow would be $1,500 - V\left(S_{ij}\right)$. As basic functions we have used the constant, S, and S^2. The option is in-the-money when $1,500 > V\left(S_{ij}\right)$.

6.4 The Case of Several Underlying Variables

6.4.1 Two GBMs: The Cholesky Factorization

Assume that we have n random variables S_i (with $i = 1, 2, \ldots, n$) each governed by the risk-neutral process:

$$\Delta S_i = rS_i\Delta t + \sigma_i S_i \sqrt{\Delta t}\varepsilon_i \tag{6.21}$$

in discrete time. Let ρ_{ik} denote the correlation coefficient between ε_i and ε_k for $1 \leq i, k \leq n$. Now a realization entails getting N random samples of ε_i (with $i = 1, 2, \ldots, n$) from a multivariate standard Normal distribution. These samples are then substituted in the individual equations Eq. (6.21) above to generate simulated paths of each S_i and to allow compute a sample value of the derivative asset.

When samples from a bivariate Normal distribution are required, a suitable procedure is the following. Take two independent samples x_1 and x_2 from a univariate $N(0, 1)$ distribution. The required samples ε_1 and ε_2 are then computed as:

$$\varepsilon_1 = x_1, \tag{6.22}$$

$$\varepsilon_2 = \rho x_1 + x_2\sqrt{1 - \rho^2}, \tag{6.23}$$

where ρ stands for the correlation coefficient between the variables of the bivariate distribution.

More in general, we can have a European option whose payoff is a function of multiple underlying assets S_1, S_2, S_d. Then we need to simulate:

$$S_i(t + \Delta t) = S_i(t)e^{\left(r - \frac{\sigma_i^2}{2}\right)\Delta t + \sigma_i\sqrt{\Delta t}\varepsilon_i}. \tag{6.24}$$

The problem is that the ε_i are correlated: $E\left[\varepsilon_i\varepsilon_j\right] = \rho_{ij}$. Here Cholesky's factorization is of great help.

Assume that we can generate d uncorrelated Normal random variables Φ_1, Φ_2, …, Φ_d. Then we can use these variables to derive correlated variables through the transformation:

$$\varepsilon = M\Phi, \tag{6.25}$$

where ε and Φ are column vectors with ε_i and Φ_i in the i-th raw. Matrix M must satisfy

$$MM^T = \Phi, \tag{6.26}$$

with Φ being the correlation matrix. This decomposition of Φ into the product of two matrices is not unique. What Cholesky's factorization does is to provide a way to choose this decomposition; in particular, it results a in matrix M which is inferior triangular.

6.4.2 Example 11: One Hundred Steps Per Year, Two GBMs

We adopt the parameter values: $T = 10$, $\Delta t = 1/100$, $r = 0.10$, $\tau_1 = 0$, and $\tau_2 = 20$. Regarding the two stochastic processes we adopt: $S_0 = 10$, $\alpha_S - \lambda_S = 0.05$, $\sigma_S = 0.40$, $I_0 = 100$, $\alpha_I - \lambda_I = 0.03$, $\sigma_I = 0.20$ and $\rho = 0.50$. We generate 100,000 realizations according to the following scheme:

$$S_{t+\Delta t} = S_t e^{\left(\alpha_S - \lambda_S - \frac{1}{2}\sigma_S^2\right)\Delta t + \sigma_S\sqrt{\Delta t}\varepsilon_t^S}, \tag{6.27}$$

$$I_{t+\Delta t} = I_t e^{\left(\alpha_I - \lambda_I - \frac{1}{2}\sigma_I^2\right)\Delta t + \sigma_I\sqrt{\Delta t}\varepsilon_t^I}. \tag{6.28}$$

At each point:

$$V\left(S_{ij}\right) = \frac{S_{ij}}{\alpha_S - \lambda_S - r}\left[e^{(\alpha_S - \lambda_S - r)\tau_2} - e^{(\alpha_S - \lambda_S - r)\tau_1}\right]. \tag{6.29}$$

With the above values: $V\left(S_{ij}\right) = 12.64S_{ij}$.

Note that the values of ε_t^S and ε_t^I must be generated with $\rho = 0.50$. To this end, as a first step we generate uncorrelated ε_t^1 and ε_t^2; then we compute:

$$\varepsilon_t^S = \varepsilon_t^1, \tag{6.30}$$

$$\varepsilon_t^I = \left[\varepsilon_t^S\rho + \varepsilon_t^2\sqrt{1-\rho^2}\right]. \tag{6.31}$$

Once we have run the simulations we have 100,000 terminal values of S, I, and V. We can compute the theoretical values of the average and the standard deviation (of S and I) using the formulas in the first example and compare them to the values derived from simulation. See Table 6.5. We further compute the correlation between the natural logarithms of the final values of S and I; we get a value 0.5031, which is very close to the $\rho = 0.50$ that we have used when generating the samples.

Next we develop the computation following LSMC. We make several changes: The option is in-the-money when $V(S_{ij}) = 12.64 S_{ij} > I_{ij}$; this feature must be added to the code for deriving the option value by LSMC. In case of investment, the resulting cash flow is $V(S_{ij}) - I_{ij} = 12.64 S_{ij} - I_{ij}$. In addition to the constant, we have used S, S^2, I, I^2, and SI as basic functions. The numerical result is 50.2262.

At the same time, if we develop again Example 11 with a binomial lattice (Sect. 4.4.2), for $T = 10$ and 12 steps per year we get a value 50.8762. And if we increase the number of steps to 100 per year then we derive a value 50.9983.

6.4.3 Example 12: European Option with a GBM and an IGBM (with Stochastic Interest Rate)

This case constitutes an extension of Example 4 above, in which we had $S_0 = 110$, $\alpha - \lambda = 0.05$, $\sigma_S = 0.40$, $T = 5$, and $K = 10$:

$$dS_t = (\alpha - \lambda) S_t dt + \sigma_S S_t dW_t^S.$$

The interest rate r follows the stochastic process:

$$dr_t = k_r (r_m - r_t) dt + \sigma_r r_t^\nu dW_t^r, \tag{6.32}$$

with $dW_t^S dW_t^r = \rho dt$. When $\nu = 0$ we have Vasicek's Model. Instead, when $\nu = 0.5$ we have CIR Model (Cox et al. 1985). Last, under the assumption $\nu = 1$ we have the IGBM Model. Now we are going to use a discount factor e^{-d_i} for each simulation path $i = 1, 2, \ldots, n$, where we construct d_i step by step according to the scheme:

$$d_i = \sum_{j=1}^{j=n} r_{ij} \Delta t = \Delta t \sum_{j=1}^{j=n} r_{ij}. \tag{6.33}$$

Table 6.5 Comparison between theoretical and simulated statistics

Variable	Statistic	Simulated	Analytic
S	Average	16.50	16.49
	Standard Dev.	33.03	32.78
I	Average	134.93	134.99
	Standard Dev.	94.64	94.67

Table 6.6 Value of call and put options when the interest rate is stochastic

σ_r	$k_r = 0.10$		$k_r = 0.20$		$k_r = 0.30$		$k_r = 0.40$	
	Call	Put	Call	Put	Call	Put	Call	Put
0.10	3.4216	1.5902	3.4102	1.5927	3.4013	1.5949	3.3942	1.5966
0.20	3.5049	1.5745	3.4808	1.5780	3.4621	1.5812	3.4473	1.5840
0.30	3.5911	1.5656	3.5533	1.5681	3.5242	1.5710	3.5014	1.5740

We further adopt the parameter values $r_0 = 0.10$, $k_r = 0.20$, $r_m = 0.10$, $\sigma_r = 0.20$, $v = 1$, and $\rho = -0.30$. At the same time that we generate r step by step we also generate the sample path for S. We run 250,000 simulations and get the option values in Table 6.6.

In Example 4, under a deterministic interest rate, the analytic solutions yielded the values $C = 3.3392$ and $P = 1.6165$. We note that the values in Table 6.6 approach them as the interest rate volatility σ_r falls. We also see that a higher reversion speed k_r dampens the effect of a stochastic interest rate.

Appendix

Below we introduce several models that can be (and have been) used for valuing investment options in different energy contexts.

A.1 Two IGBMs (One of Them a Two-Factor GBM)

Assume that we want to evaluate a base load Natural Gas Combined Cycle (NGCC) power plant. We can consider uncertain gas prices with regard to both the current level and the long-run equilibrium level. The current electricity price can be similarly assumed to be stochastic. Specifically, we assume that the three variables follow an Inhomogeneous Geometric Brownian Motion (IGBM). Thus, the time-t price of natural gas evolves according to:

$$dG_t = k_G(L_t - G_t)dt + \sigma_G G_t dW_t^G, \qquad (6.34)$$

where the long-term equilibrium vule L_t in turn follows another IGBM process:

$$dL_t = k_L(L_G - L_t)dt + \sigma_L L_t dW_t^L, \qquad (6.35)$$

with L_G acting as an anchor value. Besides, the electricity price is also an IGBM:

$$dE_t = k_E(L_E - E_t)dt + \sigma_E E_t dW_t^E. \qquad (6.36)$$

The model in a risk-neutral world would be:

$$dG_t = [k_G(L_t - G_t) - \lambda_G \sigma_G G_t]dt + \sigma_G G_t dW_t^G, \quad (6.37)$$

$$dL_t = [k_L(L_G - L_t) - \lambda_L \sigma_L L_t]dt + \sigma_L L_t dW_t^L \quad (6.38)$$

$$dE_t = [k_E(L_E - E_t) - \lambda_E \sigma_E E_t]dt + \sigma_E E_t dW_t^E. \quad (6.39)$$

We assume that $\rho_{W^G W^L} = \rho_{W^L W^E} = 0$ while $\rho_{W^G W^E} = \rho$. In addition, λ_G is the market price of risk stemming from current natural gas price (assumed to be constant); similar interpretation applies to both λ_L and λ_E.

In our Monte Carlo simulations below, we shall use the following discretization of the last three equations:

$$\Delta G_t = [k_G L_t - G_t(k_G + \lambda_G \sigma_G)]\Delta t + \sigma_G G_t \sqrt{\Delta t}\varepsilon_t^G, \quad (6.40)$$

$$\Delta L_t = [k_L L_G - L_t(k_L + \lambda_L \sigma_L)]\Delta t + \sigma_L L_t \sqrt{\Delta t}\varepsilon_t^L, \quad (6.41)$$

$$\Delta E_t = [k_E E_t - E_t(k_E + \lambda_E \sigma_E)]\Delta t + \sigma_E E_t \sqrt{\Delta t}\varepsilon_t^E. \quad (6.42)$$

The disturbances ε_t^G, ε_t^L, and ε_t^E are standard Normal variates; Δt is measured in yearly terms. Whereas ε_t^G and ε_t^L are assumed to be independent, just like ε_t^L and ε_t^E (so $\rho_{GL} = \rho_{LE} = 0$), the correlation coefficient between electricity and gas prices ρ_{GE} may be different from zero.

Regarding natural gas, it can be seen in Eqs. (6.40) and (6.41) that generating a simulation path requires knowledge of the state variable $k_G L_t$ on each day t, the three composites ($k_L L_G$, $k_G + \lambda_G \sigma_G$, $k_L + \lambda_L \sigma_L$), and the two volatilities σ_G and σ_L in the actual (as opposed to risk neutral) world.

Any simulation run fits the discretized equations Eqs. (6.40–6.42). Depending on the specific values of the correlation coefficients, the Monte Carlo simulation technique may require the generation of two or more correlated Normal variates.

The series obtained for G_t, L_t and E_t allow to compute at any time the value V_t of an investment at that time, taking into account the evolution of electricity and gas prices, as well as the behavior of the equilibrium gas price in the short term (L_t). Given the values of V_t at any moment and in each path, the LSMC approach is used. At the last moment (T), the value of the investment in each path is:

$$max(V(G_T, L_T, E_T) - I; 0) \quad (6.43)$$

At earlier moments, the method is based on the computation of a series of parameters that allow construct a linear combination of basic functions. This combination allows estimate the continuation value at each step. The specification adopted consists of a second-order expected continuation value function with 10 regressors (since there are 3 sources of risk), namely:

$$E_t^Q\left[e^{-r\Delta t}V_{t+1}(G_{t+1},L_{t+1},E_{t+1})\right] \approx a_1 + a_2G_t + a_3G_t^2 + a_4L_t + a_5L_t^2 + a_6E_t + a_7E_t^2 + a_8G_tL_t + a_9G_tE_t + a_{10}L_tE_t. \quad (6.44)$$

At any time, considering the paths that are in-the-money and by applying ordinary least squares, we can get the value of the coefficients a_1, …, a_{10}.

A.2 One GBM and Two IGBMs

Assume that we want to evaluate an investment to enhance energy efficiency in a coal-fired power plant that operates under the EU Emissions Trading Scheme (ETS). As a matter of fact, natural gas-fired power plants usually set the price in electricity markets, or their bid price is very close to the actual marginal price. Therefore, we consider three stochastic processes: natural gas price, coal price, and carbon allowance price. Gas price and carbon price contribute to determining the electricity price and therefore the expected revenues of coal-fired plants; instead, their costs are determined by coal price and carbon price.

The risk-neutral behavior of natural gas price is assumed to be governed by the following IGBM stochastic process with seasonality:

$$dG_t = df(t) + [k_G G_m - (k_G + \lambda_G)(G_t - f(t))]dt + \sigma_G(G_t - f(t))dW_t^G. \quad (6.45)$$

In this setting, G_m denotes the level to which natural gas price tends in the long run. $f(t)$ is a deterministic time function. Since we are interested in reflecting the seasonal pattern on the gas price time series throughout the year, we resort to a sinusoidal function like the cosine function: $(t) = \gamma\cos(2\pi(t+\varphi))$. Here cos stands for the cosine function measured in radians, and γ is a constant parameter (Lucia and Schwartz 2002). The cosine function has annual periodicity, hence the time is measured in years. At time $t = -\varphi$ we have $f(t = -\varphi) = \gamma$ and seasonality is highest.

Regarding coal price we adopt a stochastic process that is similar to that for natural gas but does not display seasonality:

$$dC_t = [k_C(C_m - C_t) - \lambda_C C_t]dt + \sigma_C C_t dW_t^C. \quad (6.46)$$

The notation runs akin to that for the dynamics in gas price.

The price of the emission allowance in a risk-neutral world A_t is assumed to follow a standard GBM:

$$dA_t = (\alpha - \lambda_A)A_t dt + \sigma_A A_t dW_t^A; \quad (6.47)$$

λ_A is the market price of carbon price risk.

Correlated (deseasonalised) random variables are generated according to the scheme:

$$C_{t+\Delta t} \cong \frac{k_C C_m}{k_C + \lambda_C}\left(1 - e^{-(k_C+\lambda_C)\Delta t}\right) + C_t e^{-(k_C+\lambda_C)\Delta t} + \sigma_C C_t \sqrt{\Delta t}\varepsilon_t^1, \tag{6.48}$$

$$A_{t+\Delta t} \cong A_t e^{(\alpha-\lambda_A)\Delta t} + \sigma_A A_t \sqrt{\Delta t}\left[\varepsilon_t^1 \rho_{CA} + \varepsilon_t^2 \sqrt{1-\rho_{CA}^2}\right], \tag{6.49}$$

$$\begin{aligned} G_{t+\Delta t} \cong\; & f(t+\Delta t) + \frac{k_G G_m}{k_G + \lambda_G}\left(1 - e^{-(k_G+\lambda_G)\Delta t}\right) + (G_t - f(t))e^{-(k_G+\lambda_G)\Delta t} \\ & + \sigma_G (G_t - f(t))\sqrt{\Delta t}\left[\varepsilon_t^1 \rho_{CG} + \varepsilon_t^2 \frac{\rho_{AG} - \rho_{CG}\rho_{CA}}{\sqrt{1-\rho_{CA}^2}} + \varepsilon_t^3 \sqrt{1 - \rho_{CG}^2 - \frac{(\rho_{AG} - \rho_{CG}\rho_{CA})^2}{1-\rho_{CA}^2}}\right]. \end{aligned} \tag{6.50}$$

ε_t^1, ε_t^2, and ε_t^3 are standardized Gaussian white noises with zero correlation. If samples from a standardized bivariate normal distribution are required, an appropriate procedure is the one shown above, where ρ_{GC}, ρ_{GA}, and ρ_{CA} are the correlation coefficients between the variables in the multivariate distribution.

A.3 One GBM and Two Ornstein–Uhlenbeck Processes

We could aim to study value and risk involved in coal stations operating under the EU ETS after Kyoto Protocol's expiration. We accomplish this by means of simulation techniques. Since our aim is to derive values of the Earnings at Risk (EaR), simulation must use real parameters and not risk-neutral parameters (Wilmott 2006).

We adopt the simplest mean-reverting stochastic process (also known as an Ornstein–Uhlenbeck process or O-U process) for the Clean Spark Spread (the first ingredient to the Clean Dark Spread as defined in Abadie and Chamorro 2009):

$$dS_t = k_S(S_m - S_t)dt + \sigma_S dW_t^S. \tag{6.51}$$

The current value S_t tends to the level S_m in the long term at a speed of reversion k_S. Besides, σ_S is the instantaneous volatility, and dW_t^S stands for the increment to a standard Wiener process. This model allows S_t to take on negative and positive values.

Next we adopt the notation in Kloeden and Platen (1992). The homogeneous equation is:

$$\frac{dS_t}{S_t} = -k_S dt.$$

Therefore, its fundamental solution is $\Phi_{t,t_0} = e^{-k_S(t-t_0)}$. By making $Y_t \equiv \Phi_{t,t_0}^{-1} S_t = e^{k_S(t-t_0)} S_t$, derivatives can be computed:

$$\frac{dY_t}{dS_t} = e^{k_S(t-t_0)}, \frac{d^2 Y_t}{dS_t^2} = 0, \frac{dY_t}{dt} = k_S e^{k_S(t-t_0)} S_t.$$

By Ito's Lemma:

$$dY_t = d\left(\Phi_{t,t_0}^{-1} S_t\right) = k_S S_m e^{k_S(t-t_0)} dt + e^{k_S(t-t_0)} \sigma_S dW_t^S.$$

Hence we deduce that:

$$S_t = S_0 e^{-k_S(t-t_0)} + e^{-k_S(t-t_0)} k_S \int_{t_0}^{T} e^{k_S(s-t_0)} ds + e^{-k_S(t-t_0)} \sigma_S \int_{t_0}^{T} e^{k_S(s-t_0)} dW_s^S.$$

The first moment is:

$$E(S_t) = S_0 e^{-k_S(t-t_0)} + S_m \left[1 - e^{-k_S(t-t_0)}\right]; \tag{6.52}$$

therefore: $E(S_\infty) = S_m$. The variance is given by:

$$Var(S_t) = e^{-2k_S(t-t_0)} \sigma_S^2 \int_{t_0}^{t} e^{2k_S(s-t_0)} ds = \frac{\sigma_S^2}{2k_S} \left[1 - e^{-2k_S(t-t_0)}\right]. \tag{6.53}$$

Since both mean and variance remain finite as $t \to \infty$, this process is stationary.

Equation (6.51) is the continuous-time version of a first-order autoregressive process AR(1) in discrete time:

$$S_{t+\Delta t} = S_m\left[1 - e^{-k_S \Delta t}\right] + S_t e^{-k_S \Delta t} + \varepsilon_{t+\Delta t}^S = a_S + b_S S_t + \varepsilon_{t+\Delta t}^S, \tag{6.54}$$

where $\varepsilon_{t+\Delta t}^S \sim N\left(0, \sigma_\varepsilon^S\right)$, and the following notation holds:

$$a_S \equiv S_m[1 - b_S] \to S_m = \frac{a_S}{1 - b_S}, \tag{6.55}$$

$$b_S \equiv e^{-k_S \Delta t} \to k_S = -\frac{\ln b_S}{\Delta t}. \tag{6.56}$$

Also, as shown in Abadie and Chamorro (2009, Appendix A):

$$\left(\sigma_\varepsilon^S\right)^2 = \frac{\sigma_S^2}{2k_S}\left[1 - e^{-2k_S(t-t_0)}\right] \to \sigma_S^2 = \frac{2k_S\left(\sigma_\varepsilon^S\right)^2}{1 - e^{-2k_S(t-t_0)}} = \frac{2\ln b_S\left(\sigma_\varepsilon^S\right)^2}{\Delta t\left[b_S^2 - 1\right]}. \tag{6.57}$$

Equations (6.55–6.57) will allow us to recover the continuous-time process parameters (k_S, S_m, σ_S) upon estimation of the regression coefficients (a_S, b_S) and the standard deviation of the regression residuals (σ_ε^S).

Now we turn to the second term in the Clean Dark Spread. Again we adopt an Ornstein–Uhlenbeck process for the difference $G_t/0.55 - C_t/0.40$. We have another AR(1) process as its counterpart in discrete time:

$$D_{t+\Delta t} = D_m\left[1 - \mathrm{e}^{-\mathrm{k_D}\Delta t}\right] + D_t\mathrm{e}^{-\mathrm{k_D}\Delta t} + \varepsilon^{\mathrm{D}}_{t+\Delta t} = a_{\mathrm{D}} + b_{\mathrm{D}}\mathrm{D}_t + \varepsilon^{\mathrm{D}}_{t+\Delta t}, \tag{6.58}$$

where D_t denotes the price gap at time t, D_m is the level of the gap in the long term, and $\mathrm{k_D}$ stands for the speed of reversion. The remainder of the notation goes as before.

During the current period we assume that carbon price $\mathrm{A_t}$ (in €/tCO_2) follows a GBM:

$$dA_t = \alpha A_t dt + \sigma_A A_t dW_t^A.$$

Therefore, the expected value for the allowance price in the near future is:

$$E(A_t) = A_0 e^{\alpha t}, \quad for\, t < 5. \tag{6.59}$$

At the end of this period we assume there will be a sudden jump J in price, which would push the expected value upwards:

$$t = 5^- : E(A_t) = A_0 e^{5\alpha};$$
$$t = 5^+ : E(A_t) = A_0 e^{5\alpha} + J.$$

From then on, we assume allowance scarcity is just right as an environmental policy measure and price evolves once again following a GBM:

$$E(A_t) = A_0 e^{\alpha t} + J e^{\alpha(t-5)} \quad for\, t > 5. \tag{6.60}$$

No further jumps are assumed in subsequent years for the sake of simplicity. Though environmental policy is conceivably expected to become stricter and push allowance prices to new heights at the end of this period, it is hard to foresee what will happen then.

According to Ito's Lemma, the transformed variable $X_t \equiv \ln A_t$ follows a stochastic process:

$$dX_t = \left(\alpha - \frac{\sigma_A^2}{2}\right)dt + \sigma_A dW_t^A.$$

In discrete time:

$$y_t \equiv \Delta \ln A_t = \ln A_t - \ln A_{t-\Delta t} = \left(\alpha - \frac{\sigma_A^2}{2}\right)\Delta t + \sigma_A\sqrt{\Delta t}\varepsilon_t^3, \tag{6.61}$$

where ε_t^3 is a standard Gaussian white noise.

Now, let e_1, e_2, and e_3. be uncorrelated standard normal deviates. Random samples of correlated variables can be generated as follows:

$$x_1 = f_{11} e_1, \tag{6.62}$$

$$x_2 = f_{21} e_1 + f_{22} e_2, \tag{6.63}$$

$$x_3 = f_{31} e_1 + f_{32} e_2 + f_{33} e_3, \tag{6.64}$$

where $E(x_i) = 0$, and $Cov(x_i, x_j) = \rho_{ij}$, with $i, j = 1, 2, 3, i \neq j$.

Random deviates with this correlation structure must satisfy the conditions:

$$\begin{aligned}
E(x_1^2) &= 1 \rightarrow f_{11} = 1, \\
E(x_1 x_2) &= \rho_{12} \rightarrow f_{21} = \rho_{12}, \\
E(x_2^2) &= 1 = f_{21}^2 + f_{22}^2 \rightarrow f_{22} = \sqrt{1 - \rho_{12}^2}, \\
E(x_1 x_3) &= \rho_{13} = f_{11} f_{31} \rightarrow f_{31} = \rho_{13}, \\
E(x_2 x_3) &= \rho_{23} = f_{21} f_{31} + f_{22} f_{32} \rightarrow f_{32} = \frac{\rho_{23} - \rho_{12}\rho_{13}}{\sqrt{1 - \rho_{12}^2}}, \\
E(x_3^2) &= 1 = f_{31}^2 + f_{32}^2 + f_{33}^2 = \rho_{13}^2 + \frac{\rho_{23} - \rho_{12}\rho_{13}}{1 - \rho_{12}^2} + f_{33}^2 \rightarrow \\
f_{33} &= \sqrt{1 - \rho_{13}^2 - \frac{(\rho_{23} - \rho_{12}\rho_{13})^2}{1 - \rho_{12}^2}}.
\end{aligned}$$

Therefore,

$$x_1 = e_1, \tag{6.65}$$

$$x_2 = e_1 \rho_{12} + e_2 \sqrt{1 - \rho_{12}^2}, \tag{6.66}$$

$$x_3 = e_1 \rho_{13} + e_2 \frac{\rho_{23} - \rho_{12}\rho_{13}}{\sqrt{1 - \rho_{12}^2}} + e_3 \sqrt{1 - \rho_{13}^2 - \frac{(\rho_{23} - \rho_{12}\rho_{13})^2}{1 - \rho_{12}^2}}. \tag{6.67}$$

Correlated random variables are thus generated according to the above scheme:

$$S_{t+\Delta t} = S_m \left[1 - \mathrm{e}^{-\mathrm{k_S}\Delta t}\right] + S_t \mathrm{e}^{-\mathrm{k_S}\Delta t} + \sigma_S \sqrt{\frac{1 - \mathrm{e}^{-2\mathrm{k_S}\Delta t}}{2\mathrm{k_S}}} \varepsilon_t^1, \tag{6.68}$$

$$D_{t+\Delta t} = D_m \left[1 - \mathrm{e}^{-\mathrm{k_D}\Delta t}\right] + D_t \mathrm{e}^{-\mathrm{k_D}\Delta t} + \sigma_D \sqrt{\frac{1 - \mathrm{e}^{-2\mathrm{k_D}\Delta t}}{2\mathrm{k_D}}} \left[\varepsilon_t^1 \rho_{SD} + \varepsilon_t^2 \sqrt{1 - \rho_{SD}^2}\right], \tag{6.69}$$

$$\ln A_{t+\Delta t} = \ln A_t + \left(\alpha - \frac{\sigma_A^2}{2}\right)\Delta t$$
$$+ \sigma_A \sqrt{\Delta t}\left[\varepsilon_t^1 \rho_{SA} + \varepsilon_t^2 \frac{\rho_{DA} - \rho_{SA}\rho_{SD}}{\sqrt{1-\rho_{SD}^2}} + \varepsilon_t^3 \sqrt{1 - \rho_{SA}^2 - \frac{(\rho_{DA} - \rho_{SA}\rho_{SD})^2}{1-\rho_{SD}^2}}\right], \tag{6.70}$$

where ε_t^1, ε_t^2 and ε_t^3 are standardized Gaussian white noises with zero correlation. The first expression above is derived after replacing σ_ε^S in terms of σ_S. Similarly in the second expression. At the same time, if samples from a standardized bivariate normal distribution are required, an appropriate procedure is the one shown above, where ρ_{SD}, ρ_{SA}, and ρ_{DA} are the correlation coefficients between the variables in the multivariate distribution.

References

Abadie LM, Chamorro JM (2009) Income risk of EU coal-fired power plants after Kyoto. Energ Policy 37(12):5304–5316

Brandimarte P (2002) Numerical methods in finance. Wiley, Hoboken

Broadie M, Detemple JB (2004) Option pricing: valuation models and applications. Manage Sci 50(9):1145–1177

Cox JC, Ingersoll JE, Ross SA (1985) A theory of the term structure of interest rates. Econometrica 53:385–407

Hull J (1993) Options, futures, and other derivative securities, 2nd edn. Prentice Hall, Upper Saddle River

Jäckel P (2002) Monte Carlo methods in finance. Wiley, Hoboken

Kloeden PE, Platen E (1992) Numerical solution of stochastic differential equations. Springer, Berlin

Longstaff FA, Schwartz ES (2001) Valuing American options by simulation: a simple least squares approach. Rev Financ Stud 14(1):113–147

Lucia J, Schwartz ES (2002) Electricity prices and power derivatives: evidence from the nordic power exchange. Rev Deriv Res 5(1):5–50

Wilmott P (1998) Derivatives. The theory and practice of financial engineering. Wiley, Hoboken

Wilmott P (2006) Paul Wilmott on quantitative finance. Wiley, Hoboken

Part III
Investments in the Energy Sector

It is not too much to expect that our children will enjoy in their homes electricity too cheap to meter. Lewis L. Strauss; Chairman, Atomic Energy Commission, 1954.

Chapter 7
Economic and Technical Background

7.1 Introduction

Access to energy services is an essential requirement for the development of humans and societies alike. Energy technologies bridge the gap between primary energy sources and the energy services (included transportation) that end users demand. There is a plethora of technologies for energy conversion, transmission, and distribution. Ultimately they are grounded on technological fundamentals. Nonetheless, allocation issues (what will be produced and how) along with distributive issues (who's needs are to be met) will always remain within the social realm. In other words, what is acceptable and which is the course to follow falls within social decision making.

At the time of this writing (2012–2013), the European Union has set out a road map for 2020. It draws on three main pillars: security of supply, economic efficiency, and regard for the environment. Energy efficiency plays a major role in this endeavor, since the energy that goes unconsumed need not be extracted (e.g. fossil fuels), neither transformed nor transported. Human lifestyle is thus framed in a complex system where technology and society are closely intertwined. Scientific advances, technological developments, and human values (via political institutions) set the ground for our relentless quest for higher living standards.

Below we go at some length in describing the basics of a few technologies for generating electricity. We spot important features and attributes that can serve as a basis for assessing a few specific technologies. Regarding fossil fuels, we briefly discuss some major issues of coal stations and natural gas-fired combined cycles. We also highlight key aspects of the gasification technology. Then we pay some attention to renewable energy sources; in particular, we assess an investment in a wind park. We look at their technical performance and economic assessment in each case; to this end we address the engineering elements that are potentially more relevant for economic valuation.

Figure 7.1 displays practical efficiencies for heat-to work conversion in various heat engine cycles. These efficiency levels have a major impact on the cost to producing electricity, as shown in Fig. 7.2 for a cross section of countries.

L.M. Abadie and J.M. Chamorro, *Investment in Energy Assets Under Uncertainty*, Lecture Notes in Energy 21, DOI: 10.1007/978-1-4471-5592-8_7,

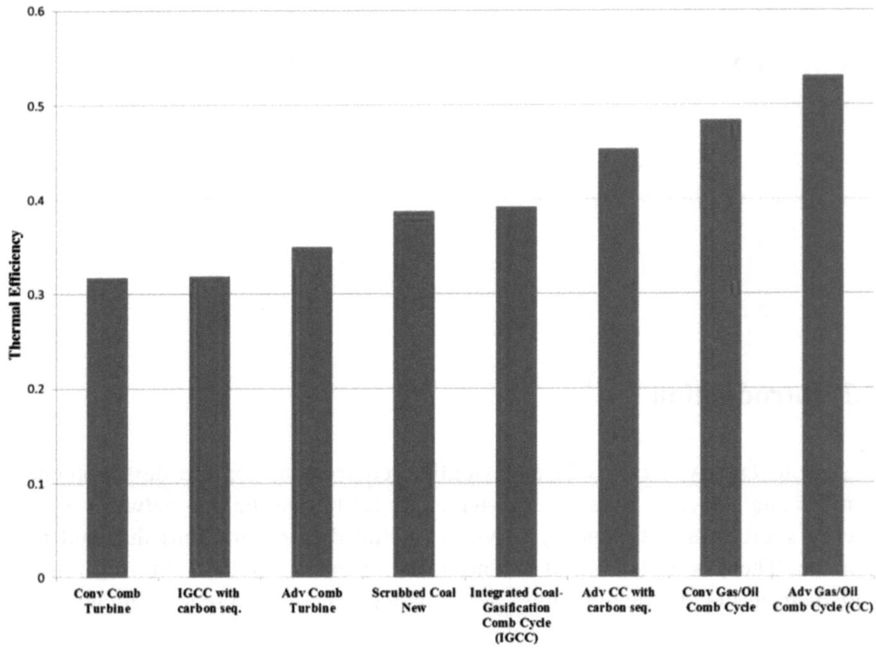

Fig. 7.1 Thermal efficiency of several power generating technologies. *Source* U.S. Energy Information Administration: Assumptions to the annual energy outlook 2012

According to Tester et al. (2005), a fossil fuel is a substance that releases energy by a chemical reaction. Most fuels release their chemical potential energy by reacting with a separate substance, namely the oxidant. For example, methane gas (the principal component of natural gas) combines with pure oxygen to form carbon dioxide and water:

$$CH_4 + 2O_2 \rightarrow CO_2 + 2H_2O. \tag{7.1}$$

The term combustion applies when fuel oxidation is rapid and produces heat primarily.

Fossil fuels are spread across the world and can be generally recovered at attractive costs; some of them, however, can only be exploited under ever costlier conditions, so a rise in demand will invariably affect their price. Besides, thanks to the fuel supply infrastructure currently available, they can be delivered to end users at affordable prices. Sometimes it is claimed that they are cheap; this can indeed be the case at first glance.

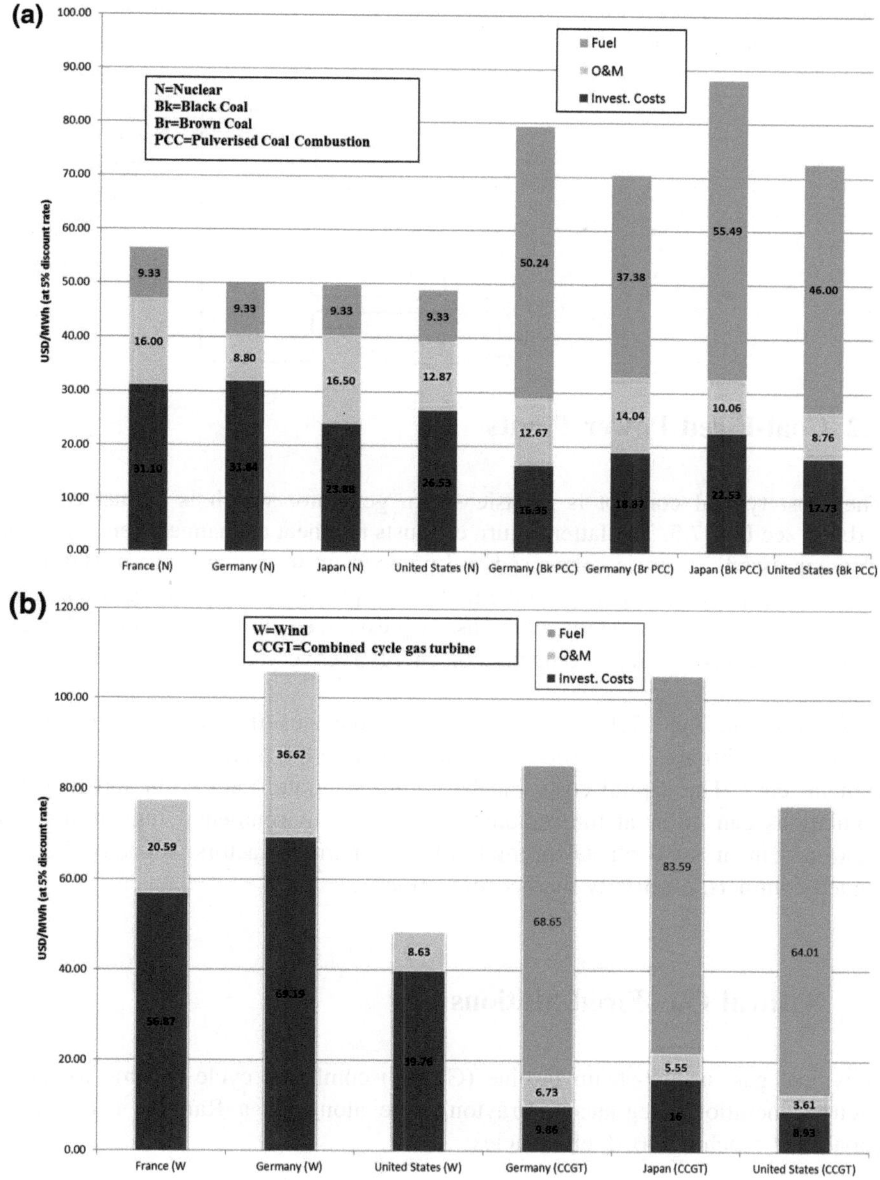

Fig. 7.2 Levelized electricity generation cost (5 % real discount rate). *Source* OECD/IEA (2010): Projected cost of generating electricity

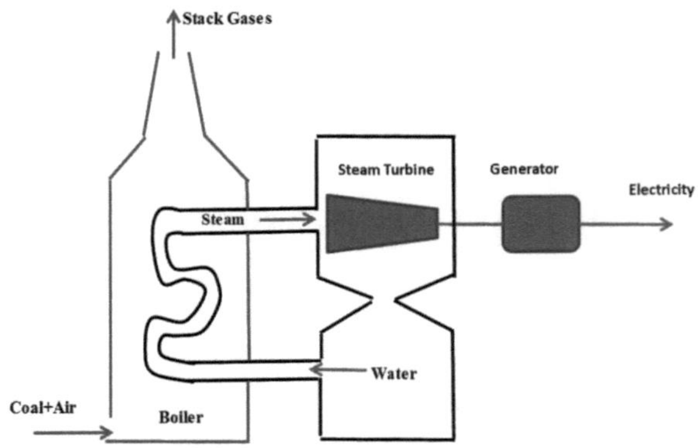

Fig. 7.3 Simple concept of a coal-fired steam generator

7.2 Coal-Fired Power Plants

The most typical concept is a basic steam generator which is connected to a turbine; see Fig. 7.3. The latter in turn exhausts to a heat exchanger; here, spent or 'decompressed' steam is condensed and pumped up to pressure again. Based on the lower heating value (LHV) of the fuel,[1] such plants reach operating efficiencies around 35 %.[2] More modern plants use a pressurized fluid bed-combined cycle design. They place a greater emphasis on energy recovery, and achieve efficiencies about 50 %.

The costs in Table 7.1 are more or less reliable; nonetheless, they can depend on local conditions (work force) or the evolution of market prices (for steel, cement, etc.). The capital costs can be highly accurate. Fuel costs and fuel use regulations can differ at the regional level. And environmental regulations can make a dent in fossil plants' profitability. Other major factors: capacity factors, signed contracts, electricity market regulation, …

7.3 Natural Gas-Fired Stations

A typical gas turbine-steam turbine (GT-ST) combined cycle scheme for electricity generation integrates a Brayton cycle alongside a Rankine cycle. It is capable of reaching 60 % efficiencies.

[1] Tester et al. (2005, p. 301): The heating value of a fuel is the maximum amount of energy obtainable by combusting one unit weight or volume of fuel and then cooling all the products of combustion (among them, water) to room temperature. If a tabulated heating value does include the energy released by condensation of the water of combustion, it is called the gross or higher heating value (HHV). Similarly, if the value does not include this latent heat, then it is called the net or lower heating value (LHV).

[2] This is a narrow definition of efficiency, as opposed to that from a systems perspective. See Connors et al. (2004).

Table 7.1 Descriptive statistics of reference power technologies

Technology	Online year	Size (MW)	Lead time (y)	Overnight cost (2010 $/kW)	Variable O & M (2010 $/MWh)	Fixed O & M (2010 $/kW)	Heat rate 2011 (Btu/KWh)	*n*th-of-a-kind heat rate (Btu/KWh)
Scrubbed coal new	2015	1,300	4	2,844	4.25	29.67	8,800	8,740
Integrated coal-gasification combined cycle (IGCC)	2015	1,200	4	3,220	6.87	48.9	8,700	7,450
IGCC with carbon sequestration	2017	520	4	5,348	8.04	69.3	10,700	8,307
Conventional gas/oil combined cycle	2014	540	3	977	3.43	14.39	7,050	6,800
Advanced gas/oil combined cycle (CC)	2014	400	3	1,003	3.11	14.62	6,430	6,333
Advanced CC with carbon sequestration	2017	340	3	2,060	6.45	30.25	7,525	7,493
Conventional combined turbine	2013	85	2	974	14.7	6.98	10,745	10,450
Advanced combined turbine	2013	210	2	666	9.87	6.7	9,750	8,550
Advanced nuclear	2017	2,236	6	5,335	2.04	88.75	10,460	10,460
Conventional hydropower	2015	500	4	2,347	2.55	14.27	9,760	9,760
Wind	2011	100	3	2,437	0	28.07	9,760	9,760
Wind offshore	2015	400	4	5,974	0	53.33	9,760	9,760

Source U.S. Energy Information Administration. Assumptions to the annual energy outlook 2012

Table 7.2 Computation of relevant parameters for techno-economic appraisal

Technology	Heat rate 2011 (GJ/MWh) (HR)	Thermal efficiency	Assumed CO_2 emission factor ($kgCO_2$/GJ)	Carbon emissions ($kgCO_2$/MWh)
Scrubbed coal new	9.2845	0.3877	94.6	878
Integrated coal-gasification combined cycle (IGCC)	9.1790	0.3922	94.6	868
IGCC with carbon sequestration	11.2891	0.3189	94.6	1,068
Conventional gas/oil combined cycle	7.4381	0.4840	56.1	417
Advanced gas/oil combined cycle (CC)	6.7840	0.5307	56.1	381
Advanced CC with carbon sequestration	7.9393	0.4534	56.1	445
Conventional combined turbine	11.3366	0.3176	56.1	636
Advanced combined turbine	10.2868	0.3500	56.1	577

Using these estimates we can compute a number of relevant input and output parameters. See Table 7.2. We assume that the capacity factor is 80 %; thus the station operates 7,008 h over the year (=365 × 24 × 0.80).

According to IPCC (2006), a plant burning natural gas has a carbon emissions factor of 56.1 $kgCO_2$/GJ. Since under 100 % efficiency conditions 3.6 GJ would be consumed per megawatt-hour (i.e. $1GJ = 1/3.6$ MWh), we get

$$I_G = \frac{0.20196}{E_G} \frac{tCO_2}{\text{MWh}}, \tag{7.2}$$

where I_G stands for the emission intensity of the plant (tCO_2/MWh), which in turn depends on the net thermal efficiency of each gas-fired plant, E_G.

Similarly, following IPCC (2006) a plant burning bituminous coal has an emission factor of 94.6 $kgCO_2$/GJ under 100 % efficiency conditions; then,

$$I_C = \frac{0.34056}{E_C} \frac{tCO_2}{\text{MWh}}. \tag{7.3}$$

Two different spreads can be defined. First, for a gas-fired plant the clean spark spread (*CSS*) is

$$CSS = P_E - \frac{P_G}{E_G} - P_{CO_2} I_G, \tag{7.4}$$

where P_E denotes electricity price ($/MWh), P_G is the price of natural gas ($/MWh), and P_{CO_2} is the price of a EU emission allowance ($/$tCO_2$).[3] Thus the complete formula for the *CSS* is:

[3] 1 MWh = 3.412 mmBTU; alternatively, 1 mmBTU = 0.293083 MWh.

$$CSS = P_E - \frac{1}{E_G}(P_G + 0.20196 \times P_{CO_2}). \tag{7.5}$$

Analogously, the complete formula for the Clean Dark Spread (*CDS*) of a coal-fired station is:

$$CDS = P_E - \frac{1}{E_C}(P_C + 0.34056 \times P_{CO_2}), \tag{7.6}$$

where P_C is the price of coal ($/MWh).

We can solve for P_E in Eq. (7.5) and then substitute into Eq. (7.6), thus linking the two spreads. This yields:

$$CDS = CSS + \frac{1}{E_G}(P_G + 0.20196 \times P_{CO_2}) - \frac{1}{E_C}(P_C + 0.34056 \times P_{CO_2}). \tag{7.7}$$

Or, equivalently,

$$CDS = CSS + \left(\frac{P_G}{E_G} - \frac{P_C}{E_C}\right) + \left(\frac{0.20196 \times P_{CO_2}}{E_G} - \frac{0.34056 \times P_{CO_2}}{E_C}\right). \tag{7.8}$$

Regarding the right-hand side, the first term represents the margin derived by natural gas plants. The second one is the fuel price gap (adjusted for relative efficiency rates). To the extent that this gap is positive, it is to the advantage of coal-fired plants. The last term, though, stands for the disadvantage of coal plants because of their higher carbon emissions, which is further aggravated by their lower efficiency levels.

Both spreads are equal ($CSS = CDS$) when

$$\left(\frac{P_G}{E_G} - \frac{P_C}{E_C}\right) = \left(\frac{0.34056 \times P_{CO_2}}{E_C} - \frac{0.20196 \times P_{CO_2}}{E_G}\right), \tag{7.9}$$

i.e., when the advantage in fuel price is wiped out by the higher emission costs.

For high enough allowance prices the *CDS* can reach a low value or even become negative. Specifically, we would get $CDS = 0$ when

$$CSS + \left(\frac{P_G}{E_G} - \frac{P_C}{E_C}\right) = \frac{0.34056 \times P_{CO_2}}{E_C} - \frac{0.20196 \times P_{CO_2}}{E_G}, \tag{7.10}$$

i.e., when the higher costs of carbon permits offset not only the fuel price gap but the *CSS* as well.

An important point to be considered in many deregulated electricity markets is that gas-fired plants are the marginal units that set the price of electricity. This fact not only affects their profit margins, but those of coal stations as well.

7.4 Gasification Plants

Some naturally occurring fossil fuels can undergo a chemical or physical transformation to improve their quality. Petroleum refining and coal gasification are examples of complex fuel conversion processes. Similarly, natural gas can be converted to liquid transportation fuels, like diesel and gasoline. Two basic mechanisms of fuel conversion are rejection of carbon and addition of hydrogen.

Coal in particular can be subject to thermal treatment (pyrolysis) and converted to various alternative fuels. For example, coal liquefaction allows produce a substitute for petroleum-derived, consumer-grade liquid fuel. Coal gasification, instead, produces ('synthetic') gas (e.g. hydrogen or methane) from coal. Historically, growing concerns about depletion of domestic supplies and greater awareness of environmental issues have provided a big push to private and public programs aiming at cheaper and cleaner fuels.

7.5 Wind Parks

Humans have used renewable energies for centuries to provide energy services. Despite the long history, however, they only provide a small fraction of the world's primary energy. They rank high in terms of environmental benefits yet not so high regarding short-term economics. Figure 7.4 shows the evolution of wind power capacity installed worldwide. Figure 7.5, instead, refers to that of solar power.

Winds are a particular type of solar energy in that they arise from uneven solar heating of land and sea surfaces. The potential for wind seems to be quite large; indeed, well beyond global needs. Marvel et al. (2012) use a climate model to estimate the amount of power that can be extracted from both surface and high-altitude winds, considering only geophysical limits. According to their results, surface wind turbines alone could extract kinetic energy at a rate of at least 400 TW (TW, one trillion watts) while the level of present global primary power demand approaches 18 TW. On the other hand, Jacobson and Archer (2012) define the saturation wind power potential as the maximum wind power that can be extracted upon increasing the number of wind turbines over a large geographic region, independent of societal, environmental, climatic, or economic considerations. This saturation potential is over 250 TW at 100 m up globally (100 m above ground is the hub height of most modern wind turbines), assuming conventional wind turbines distributed everywhere on Earth.

According to Tester et al. (2005), it is standard practice to estimate the maximum efficiency attainable by a wind turbine using an ideal, somewhat oversimplified, fluid flow model. The so-called Betz limit turns out to be 0.593. Thus, gross power coefficients of 50 % can be achieved by modern turbines; then, gearbox and electrical losses involve a net value of 40 %. Note, though, this limit

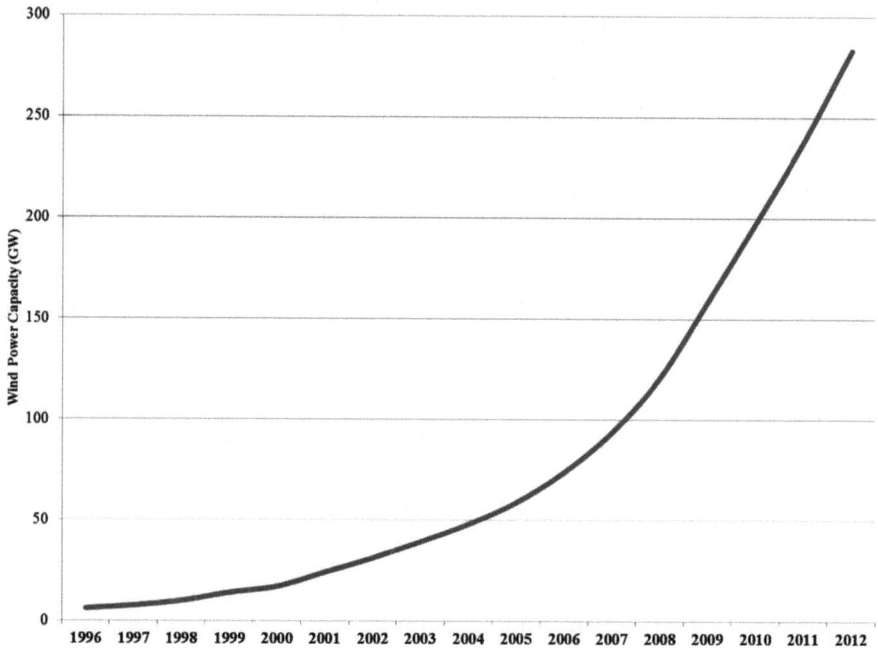

Fig. 7.4 Time evolution of wind power capacity installed worldwide. *Source* Ren21 (2013): Renewables 2013 global status report

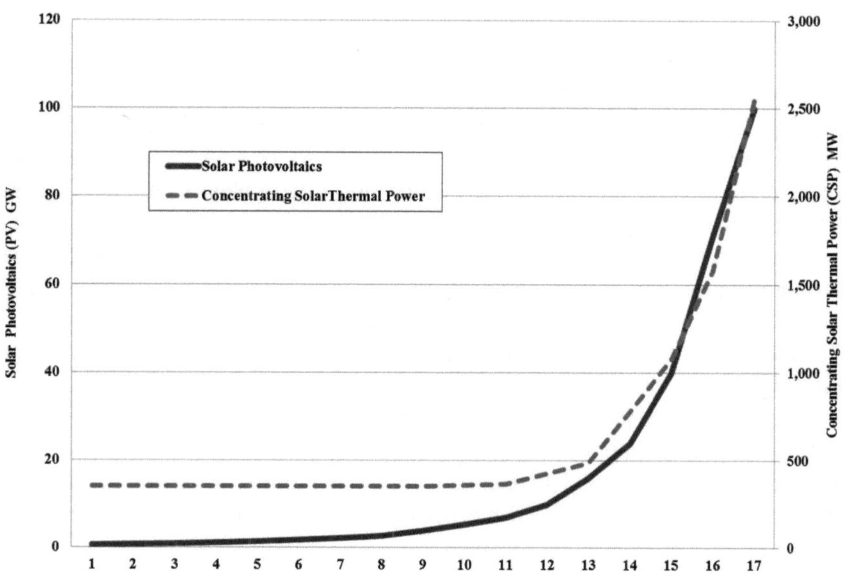

Fig. 7.5 Time evolution of solar power capacity installed worldwide. *Source* Ren21 (2013): Renewables 2013 global status report

is only as good as the underlying simplified model; it must be taken as a useful first-order approximation to reality.

Now, public support to renewable energies is usually justified on three grounds: climate change, security of supply, and industrial policy. Some of the positive effects from renewables' development are global, e.g. the abatement of greenhouse gas emissions, and the reduction of investment unit costs (because of the learning effect). Impacts from enhanced energy security and industrial policy, instead, are derived at the national level.

Renewable sources are getting ever more relevant in the generation of electric energy. Major drivers are the decreasing costs of renewable technologies and strong support from government agencies. This trend is expected to continue in the years ahead (European Commission 2011). Pérez-Arriaga and Batlle (2012) analyze the impact of a strong penetration of renewable, intermittent generation on the planning, operation, and control of power systems. See also EWEA (2010) and NREL (2010).

Within this set of technologies wind stands apart, with solar photovoltaic (PV) and concentrated solar power (CSP) somewhat behind. The increasing role of these intermittent generation technologies gives rise to important challenges in the operation of the electric system. Regarding solar energy, it is more predictable than wind over short periods of time. It also displays a diurnal seasonality which overlaps with the hours of strongest load thus coinciding with the times of highest prices. This suggests that the prices at the times of strongest operation of solar plants will approach peak prices.

Despite its potential, empirical evidence shows that actual deployment of wind energy lags far behind that potential. Some physical constraints on its use include intermittence, uncertainty, seasonality, non-dispatchability, and distance from load pockets. The problem posed by intermittence, however, is less acute when dealing with a large balancing area since the behavior of wind correlates less than perfectly across all the sites in the area (provided there is enough transmission capacity). Further deployment of renewable energies (wind in particular) would also benefit significantly from greater storage capacities. A minor concern is that wind energy is not quite carbon free.[4] Large-scale deployment of turbines can also disrupt local wildlife and fauna, affect local temperature and even global weather. These negative impacts are hard to quantify but this does not render them less real.

But most probably this is not the whole story. Several barriers (whether economic, social, or other type) are probably playing a role in hampering adoption across the globe. Regarding economic barriers, casual observation allows to identify a number of support schemes which are presumably aimed at providing greater certainty to potential investors in this technology; see Klessmann et al. (2008). In other words, uncertain returns on these investments are considered a

[4] For example, the very construction of a wind turbine consumes energy (fossil to a large extent). Ortegon et al. (2012) report a CO_2 emission factor for wind power in the range 20–38 and 9–13 gCO_2/kWh for on-shore and offshore applications, respectively. Of course, this consideration also applies to coal stations or nuclear plants.

major cause for concern both to developers and investors alike (alongside others like electricity grid- and market-related barriers).

Actual support programs typically rely on a combination of different measures such as special tax regimes, cash grants, or financial incentives; an overview can be found in Daim et al. (2012) and Snyder and Kaiser (2009). So-called Renewable Energy Feed-in Tariffs (REFIT) are a guaranteed payment to generators of renewable electricity (say, 90 €/MWh, for instance) over a certain period of time (e.g. 20 years). This instrument is typical in several EU countries, among them Germany. Spain allows similarly this remuneration option. Nonetheless, wind power generators seem to prefer the alternative option, namely a premium on top of the electricity market price. The UK instead incentivizes renewable electricity through the use of renewable energy credits (the Renewables Obligation Certificates, or ROCs) which are further traded in their specific market. EU nations also grant some tax exemptions (for instance, from carbon taxes) and subsidies (to capital expenditure). In the US there is a production tax credit at the federal level. The fact that it has expired three times over the last ten years is not reassuring, however. A number of States have set renewable portfolio standards whereby a certain fraction of the State's electricity must come from renewable sources. Some States also take part in a regional greenhouse gas initiative, a cap-and-trade market for carbon. Regarding subsidies, they are both lower and less certain than those in Europe.

A suitable valuation approach for wind projects must not only account for intermittence and uncertainty. It must also take account of their irreversible character and the flexibility enjoyed by project managers (e.g. the option to delay investment). Under these circumstances, traditional valuation techniques based on discounted cash flows have been found inferior to contingent claims or real options analysis.

Following the latter approach, Boomsma et al. (2012) assess both the time and the size of the investment in renewable energy projects under different support schemes. They consider up to three sources of uncertainty: steel price, electricity price, and subsidy payment, all of which are assumed to follow uncorrelated geometric Brownian motions (with the last one modulated by Markov switching). For illustration purposes, they focus on a Norwegian case study. According to their results, a fixed feed-in tariff encourages earlier investment than renewable energy certificates. The latter, though, create incentives for larger projects.

Reuter et al. (2012) instead pick Germany as a case study. In their model the electric utility decides whether to add new generation capacity or not once a year over the planning horizon. The new capacity can be either a fossil fuel power plant (with a constant load factor) or a wind power plant (with a normally distributed load factor), both equally sized. The yearly electricity price is subject to (normally distributed) exogenous shocks (assumed independent from wind load factor). The third source of uncertainty concerns climate policy; it is represented by the feed-in tariff which is a Markov chain with two possible values and a given transmission matrix. This risk factor is also assumed independent from the other two. Their results stress the importance of explicitly modeling the variability of renewable loads owing to their impact on profit distributions and the value of the firm. Besides, greater uncertainty about the future behavior of the feed-in tariff requires much higher trigger tariffs for which renewable investments become attractive (i.e. equally profitable as a coal-fired station of equal capacity).

Abadie and Chamorro (2012) address the present value of an investment in a wind park and the optimal time to invest under different payment settings: (a) A fixed feed-in tariff for renewable electricity over 20 years of useful life. (b) Electricity price as determined by the market. (c) A combination of the market price and a constant premium. (d) A transitory subsidy available only at the initial time. We also develop sensitivity analyses with respect to changes in the investment option's maturity and electricity price volatility.

Our paper differs from others in several respects. We consider two sources of uncertainty. We assume more general stochastic processes for the state variables; in particular, we account for mean reversion in commodity prices. We develop a trinomial lattice that supports this behavior. We also make room for seasonal behavior in the price of electricity and in wind load factor. Indeed, they turn out to be correlated to some degree, which has been typically overlooked despite its impact on project value. The underlying dynamics in the price of electricity is estimated from observed futures contracts with the longest maturities available (namely, up to five years into the future); this includes the market price of electricity price risk. The dynamics of wind load factor is also estimated from actual (monthly) time series alongside seasonality. The riskless interest rate is also taken from (financial) markets. Both the project's life and the option's maturity are finite; in our simulations below the size of the time step is not $\Delta t = 1$ (or one step per year), but a much shorter $\Delta t = 1/60$ (five steps per month). In addition to a fixed feed-in tariff and a premium over electricity price, another support scheme that we consider is an investment subsidy that is only available at the initial time but is foregone otherwise. We further provide numerical estimates of the trigger investment cost below which it is optimal to invest immediately.

7.6 Futures Markets

Below we describe the defining characteristics of several futures contracts. We draw on these contracts when assessing some investment opportunities in the next chapters.

The ICE Brent Crude futures contract is a deliverable contract based on exchange of futures for physical (EFP) delivery with an option to cash settle. The West Texas Intermediate Light Sweet Crude Oil futures contract is cash settled against the prevailing market price for US light sweet crude.

ICE UK Base Electricity Futures Contracts are for physical delivery of Electricity on a continuous baseload basis, i.e. 23:00–22:59 LLT Monday–Sunday, through National Grid, the transmission system operator in the UK. Delivery is made equally each hour throughout the delivery period.

ICE UK Natural Gas Futures Contracts are for physical delivery through the transfer of rights in respect of Natural Gas at the National Balancing Point (NBP) Virtual Trading Point, operated by National Grid, the transmission system operator in the UK. Delivery is made equally each day throughout the delivery period. There are 78–83 consecutive month contracts.

The ICE Rotterdam Coal Futures Contract is cash-settled at an amount equal to the monthly average API 2 Index as published in Argus/McCloskey's Coal Price Index Report. There are 61–72 consecutive month contracts.

The ICE ECX EUA Futures Contract is a deliverable contract where each Clearing Member with a position open at cessation of trading for a contract month is obliged to make or take delivery of emission allowances to or from National Registries in accordance with the ICE Futures Europe Regulations. Table 7.3 provides some further details on these contracts.

On January 8th 2013 the contracts in Table 7.4 were traded on their respective markets.

Table 7.3 Details about futures contracts on particular energy commodities

Futures contract	Quotation	Contract size
ICE brent crude	USD and cents per barrel[1]	1,000 barrels
ICE WTI crude	USD and cents per barrel	1,000 barrels
ICE UK base electricity	Sterling and pence per MWh	1 MWh of electricity per hour per day
ICE UK natural gas	Sterling and pence per therm[2]	1,000 therms per day per delivery period
ICE rotterdam coal	US dollar and cents per tonne	1,000 metric tonnes of thermal coal
ICE ECX EUA futures	Euro and cents per metric tone	1,000 CO_2 EU allowances

[1] 1 barrel = 42 US gallons
[2] 1 therm = 29.3071 kWh

Table 7.4 Futures contracts traded on 01/08/2013

Futures contract	Number of contracts
ICE brent crude	53 monthly contracts (Feb-13–Dec-19)[1]
ICE WTI crude	77 monthly contracts (Feb-13–Dec-21)[2]
ICE UK base electricity	56 monthly contracts (Feb-13–Sep-17)
ICE UK natural gas	80 monthly contracts (Feb-13–Sep-19)
ICE rotterdam coal	72 monthly contracts (Jan-13–Dec-18)
ICE ECX EUA futures	December maturities (Dec-13–Dec-20)[3]

[1] From Dec-16 to Dec-19 only contracts with maturity June and December
[2] From Dec-18 to Dec-21 only contracts with maturity June and December
[3] Contracts are listed on a quarterly expiry cycle such that March, June, September and December contract months are listed up to June 2013 and annual contracts with December expiries for 2013 up to 2020

References

Abadie LM, Chamorro JM (2012) Valuation of wind energy projects: a real options approach. Basque centre for climate change (bc3) working paper 2012-11

Boomsma TK, Meade N, Fleten S-T (2012) Renewable energy investments under different support schemes: A real option approach. Eur J Oper Res 220:225–237

Connors S, Martin K, Adams M, Kern E (2004) Future electricity supplies: redefining efficiency from a systems perspective. LFEE-WP-04-005

Daim TU, Amer M, Brenden R (2012) Technology roadmapping for wind energy: case of the Pacific Northwest. J Clean Prod 20(1):27–37

European Commission (2011) Communication from the Commission to the European Parliament, the Council, the European Economic and Social Committee and the Committee of Regions. Energy Roadmap 2050. Belgium. http://ec.europa.eu/energy/energy2020/roadmap/index_en.htm

European Wind Energy Association (2010) Wind Barriers: Administrative and grid access barriers to wind power. Brussels, Belgium. http://www.windbarriers.eu/fileadmin/WB_docs/documents/WindBarriers_report.pdf

IPCC (2006) Guidelines for national greenhouse gas inventories

Jacobson MZ, Archer CL (2012) Saturation wind power potential and its implications for wind energy. PNAS 109(39):15679–15684

Klessmann C, Nabe C, Burges K (2008) Pros and cons of exposing renewables to electricity market risks—A comparison of the market integration approaches in Germany, Spain, and the UK. Energy Policy 36:3646–3661

Marvel K, Kravitz B, Caldeira C, (2012) Geophysical limits to global wind power. Nature Climate Change, published on line 9 September

National Renewable Energy Laboratory (2010) Western wind and solar integration study. Prepared by GE Energy. http://www.nrel.gov/docs/fy10osti/47781.pdf

OECD/IEA (2010) Projected cost of generating electricity

Ortegon K, Nies LF, Sutherland JW (2012) Preparing for end of service life of wind turbines. J Cleaner Prod 39:191–199

Pérez-Arriaga IJ, Batlle C (2012) Impacts of intermittent renewables on electricity generation system operation. Economics of Energy & Environmental Policy 1(2):3–17

Ren21 (2013) Renewables 2013 global status report

Reuter WH, Szolgayová J, Fuss S, Obersteiner M (2012) Renewable energy investment: Policy and market impacts. Appl Energy 97:249–254

Snyder B, Kaiser MJ (2009) A comparison of offshore wind power development in Europe and the US: Patterns and drivers of development. Appl Energy 86(10):1845–1856

Tester JW, Drake EM, Driscoll MJ, Golay MW, Peters WA (2005) Sustainable Energy: Choosing among options. The MIT Press

Chapter 8
Valuation of Energy Assets: A Single Risk Factor

8.1 Introduction

We introduce some simple valuation examples. They are meant to be just a first approach to the valuation methods under uncertainty that are applied to energy assets. In other words, more complex models would probably be required in most of the cases. Anyway, whatever the particular application at hand, the valuation model draws on numerical estimates of the relevant parameters (whether they refer to deterministic variables or stochastic processes). Table 8.1 shows these values for each commodity (denoted by i); they will be used throughout in the valuations that follow.

8.2 Case 1: An Advanced Gas/Oil Combined Cycle

Here the price of natural gas is assumed stochastic. Electricity price, instead, is deterministic ($\sigma_E = 0$); it evolves according to an average trend (as estimated from futures markets). Admittedly, this example may not be very realistic. In a number of deregulated electricity markets, gas stations turn out to be the marginal units that set the price of electricity. As a consequence, natural gas and electricity prices are highly correlated. Nonetheless, this example aims to provide a simpler introduction. In the next chapters the valuation will account for two or more stochastic processes. For the time being, the NPV here derived is correct (as long as its computation involves the futures curves on electricity and natural gas), provided no other major driver is absent (e.g. the carbon emission allowance price in certain markets). Table 8.2 displays the parameter values of the gas station; they are taken from Tables 7.1 and 7.2. They are necessary for computing the set of parameters that are shown in Table 8.3. We consider that the plant has a useful life of 25 years.

Assuming that all O&M costs grow at the risk-free interest rate ($r = 2.05$ %), the present value of all deterministic costs amounts to $I = 765.35$ M\$. At the same

L.M. Abadie and J.M. Chamorro, *Investment in Energy Assets Under Uncertainty*, Lecture Notes in Energy 21, DOI: 10.1007/978-1-4471-5592-8_8,

Table 8.1 Parameter values of the underlying stochastic commodity price processes

Parameter	Electricity ($/MWh)	Natural gas ($/MWh)	Oil ($/barrel)	Coal ($/t)	CO_2 ($/t)
$S_0^i - f_i(0)$	65	30	110	90	10
$k^i S_m^i / k^i + \lambda^i$	150	50	90	140	–
$k^i + \lambda^i$	0.06	20	0.30	0.20	−0.05

Table 8.2 Basic parameters of the natural gas-fired power plant

Size (MW)	400	Variable O&M cost ($/MWh)	3.11
Lead time (years)	3	Fixed O&M cost ($/kW)	14.62
Overnight cost ($/kW)	1,003	Heat rate (Btu/kWh)	6,430
Capacity factor (%)	80	Thermal efficiency	0.5307

Table 8.3 Resulting parameters for the natural gas-fired power plant

Yearly output (MWh)	2,803,200	Variable O&M cost (M$/year)	8.718
Heat rate (GJ/MWh)	6.7840	Fixed O&M cost (M$/year)	5.848
Overnight cost (M$)	401.20	CO_2 emissions (tCO_2/year)	1,066,898

time, the present value of the natural gas consumed is 4,646.09 M$, while that of the electricity produced is 5,764.74 M$. Therefore, we get $NPV = 353.31$ M$. If the capacity factor were as low as 48.62 % this would yield $NPV = 0$; below that operation level, we would get $NPV < 0$.

Now we address the American option to invest in a gas station. As time goes on, we face a higher (deterministic) electricity price, whereas the gas price changes stochastically. The starting price of electricity at time t is given by the following equation:

$$S_t^E = \frac{k^E S_m^E}{k^E + \lambda^E}\left(1 - e^{-\left(k^E + \lambda^E\right)t}\right) + S_0^E e^{-\left(k^E + \lambda^E\right)t}. \tag{8.1}$$

The initial price of natural gas, S_t^G, instead, will depend on the levels reached on the nodes of the binomial lattice.

Next we compute the present value of an annuity yielding 1 MWh of electricity per year over 25 years (from date $\tau_1 = t + 3$ to $\tau_2 = t + 28$). This value VA_t^E is a function of the initial price S_t^E:

$$VA_t^E\left(S_t^E\right) = \frac{S_t^E - \frac{k^E S_m^E}{k^E + \lambda^E}}{k^E + \lambda^E + r}\left[e^{-(k^E + \lambda^E + r)\tau_1} - e^{-(k^E + \lambda^E + r)\tau_2}\right] + \frac{k^E S_m^E}{r(k^E + \lambda^E)}\left[e^{-r\tau_1} - e^{-r\tau_2}\right]. \tag{8.2}$$

Similarly, the present value of an annuity yielding 1 MWh of natural gas each year between dates $\tau_1 = t + 3$ and $\tau_2 = t + 28$, denoted by VA_t^G, depends on the initial price S_t^G:

$$VA_t^G(S_t^G) = \frac{S_t^G - \frac{k^G S_m^G}{k^G + \lambda^G}}{k^G + \lambda^G + r}\left[e^{-(k^G+\lambda^G+r)\tau_1} - e^{-(k^G+\lambda^G+r)\tau_2}\right]$$

$$+ \frac{k^G S_m^G}{r(k^G + \lambda^G)}\left[e^{-r\tau_1} - e^{-r\tau_2}\right]. \quad (8.3)$$

From the above expressions, the net present value of investing in the gas station at time t (in M\$) is:

$$NPV_t(S_t^E, S_t^G) = 2.8032 VA_t^E(S_t^E) - \frac{2.8032}{0.5307} VA_t^G(S_t^G) - I. \quad (8.4)$$

Consider that the option to invest in the gas plant expires in 10 years. We develop a one-dimensional binomial lattice. The option value W at the final nodes is:

$$W_T = \max\left[NPV_T(S_T^E, S_T^G), 0\right]. \quad (8.5)$$

In this example, the electricity price at that time S_T^E is the same across all the nodes. At previous times, instead, the option value is given by:

$$W_t = \max\left[NPV_t(S_t^E, S_t^G), e^{-r\Delta t}(p_u W^+ + p_d W^-)\right]. \quad (8.6)$$

Taking 100 time steps per year we compute an option value of 888.66 M\$. Note that in this case the spark spread (SS) is growing over time. Obviously, both the value of the station and that of the option to invest in it depend on the capacity factor. Table 8.4 shows how valuations are affected by this parameter. The capacity factor plays a major role in the profitability of the plant and its appeal as an investment opportunity.

8.3 Case 2: A New Scrubbed Coal-Fired Station

Now we assess another power technology, namely a new coal station. Again, the electricity price is assumed deterministic, while that of the input fuel is stochastic. Table 8.5 shows the parameter values of the coal plant. They are used for computing those displayed in Table 8.6. The plant has a useful life of 50 years.

Consider that the plant burns coal with a calorific value of 6,000 kcal/Kg. Then, under 100 % efficiency, each tonne of coal would allow to generate 6.978 MWh of electricity. In other words, under these efficiency circumstances, generating one MWh would require 0.143308 coal tones.

Table 8.4 Impact of the capacity factor on the NPV and the option value of the gas station

Capacity f.	30 %	40 %	50 %	60 %	70 %	80 %
NPV (M$)	−209.64	−97.05	15.54	128.13	240.72	353.31
Option V	73.80	227.86	390.81	555.98	722.05	888.86

Table 8.5 Basic parameters of the scrubbed coal–fired power plant

Size (MW)	1,200	Variable O&M cost ($/MWh)	4.25
Lead time (years)	4	Fixed O&M cost ($/kW)	29.67
Overnight cost ($/kW)	2,844	Heat rate (Btu/kWh)	8,800
Capacity factor (%)	80	Thermal efficiency	0.3877

Table 8.6 Resulting parameters for the scrubbed coal station

Yearly output (MWh)	8,409,600	Variable O&M cost (M$/year)	35.7408
Heat rate (GJ/MWh)	9.2845	Fixed O&M cost (M$/year)	35.604
Overnight cost (M$)	3,412.80	CO_2 emissions (tCO2/year)	7,387,086

Assuming that O&M costs grow at the risk-free interest rate ($r = 0.02$), the present value of all deterministic costs amounts to $I = 6{,}980.04$ M$. The present value of the coal consumed is 12,404.03 M$, whereas that of the electricity produced reaches 30,434.38 M$. Therefore, we get $NPV = 11{,}050.31$ M$. The NPV would fall to zero if the capacity factor fell to 25.58 %; below this level, we would get $NPV < 0$. Note that here the utility faces no cost from carbon emission allowances.

The better results of the coal plant, relative to those of the gas station, owe to:

(a) The installed capacity of the former is 3 times that of the latter.
(b) The coal plant operates for 50 years, while the gas station works half that time.
(c) The cost of emission allowances is left apart.
(d) The expectation of an increasing electricity price, which would rise at a higher rate than coal price, with a stronger impact in the long run.

Now we address an American option to invest in a coal-fired plant. The present value of an annuity yielding 1 MWh of electricity per year over 50 years (from date $\tau_1 = t + 4$ to $\tau_2 = t + 54$) is formally the same as in the case of the gas plant:

$$VA_t^E\left(S_t^E\right) = \frac{S_t^E - \frac{k^E S_m^E}{k^E + \lambda^E}}{k^E + \lambda^E + r}\left[e^{-(k^E+\lambda^E+r)\tau_1} - e^{-(k^E+\lambda^E+r)\tau_2}\right] + \frac{k^E S_m^E}{r(k^E + \lambda^E)}\left[e^{-r\tau_1} - e^{-r\tau_2}\right]. \tag{8.7}$$

This value VA_t^E is a function of the initial price S_t^E, which is given by Eq. (8.1). Similarly, the present value of an annuity yielding 1 t of coal each year between dates $\tau_1 = t + 4$ and $\tau_2 = t + 54$, denoted by VA_t^C, depends on the initial price S_t^C:

$$VA_t^C(S_t^C) = \frac{S_t^C - \frac{k^C S_m^C}{k^C+\lambda^C}}{k^C + \lambda^C + r}\left[e^{-(k^C+\lambda^C+r)\tau_1} - e^{-(k^C+\lambda^C+r)\tau_2}\right]$$

$$+ \frac{k^C S_m^C}{r(k^C + \lambda^C)}\left[e^{-r\tau_1} - e^{-r\tau_2}\right]. \tag{8.8}$$

From the above expressions, the net present value (in M$) of investing in the coal plant at time t is:

$$NPV_t(S_t^E, S_t^C) = 8.4096\ VA_t^E(S_t^E) - \frac{8.4096}{0.3877 \times 6.978} VA_t^C(S_t^C) - I. \tag{8.9}$$

We develop a one-dimensional binomial lattice. The option value W at the final nodes is:

$$W_T = \max\left[NPV_T(S_T^E, S_T^C), 0\right]. \tag{8.10}$$

In this example, the electricity price at that time S_T^E is the same across all the nodes. At previous times, instead, the option value is given by:

$$W_t = \max\left[NPV_t(S_t^E, S_t^C), e^{-r\Delta t}(p_u W^+ + p_d W^-)\right]. \tag{8.11}$$

Following this procedure we compute W for different times to maturity. The results appear in Table 8.7. As usual, when there is no option to wait, $W = NPV$; this amounts to 11,050 M$. When $T = 10$, however, the option value rises to 11,370 M$. Fig. 8.1 displays the results.

Next we consider two alternative investment opportunities. The first one involves the construction of a coal station immediately (at $t = 0$). The plant takes 4 years to build and runs for 50 years. Therefore, it would operate from $t = 4$ to $t = 54$. This option to invest is assessed against a (sequential) alternative project involving natural gas:

(a) The European option to build three natural gas plants each 400 MW in one year. Thus, decision would take place at $t = 1$, they would start operation at $t = 4$, and would work until $t = 29$.
(b) Plus the European option to build another three natural gas plants, each 400 MW, 25 years later. Thus, decision would take place at $t = 26$, they would start operation at $t = 29$, and would work until $t = 54$. Thus, both the coal station and the gas plant fleet would cease operation at the same time.

Table 8.8 shows the NPV of the two alternatives. According to it, the coal station is the preferred investment by far. Note the implicit assumption of no carbon constraints in this example. Similarly, the second vintage of gas plants could be cheaper if, for example, the utility already holds the property right over the site.

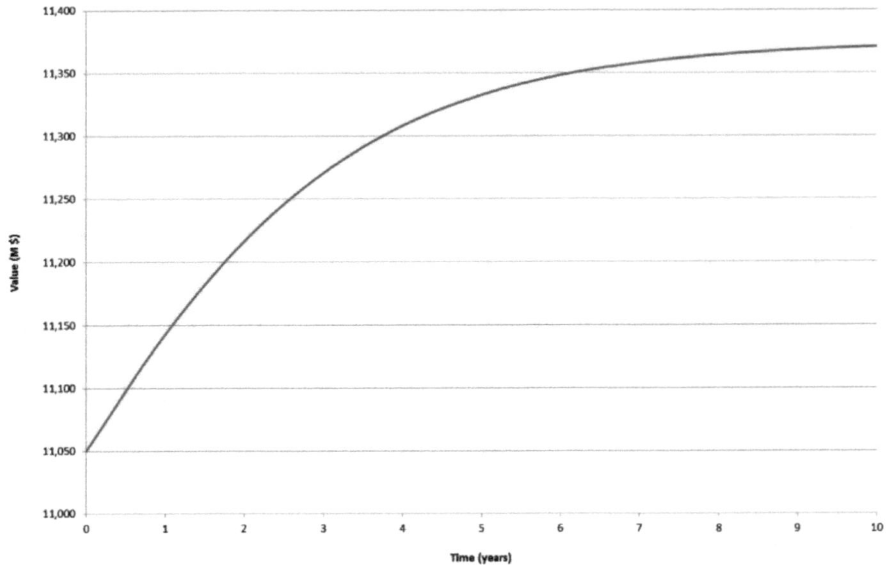

Fig. 8.1 Option value (M$) as a function of the time to maturity (years)

Table 8.7 Option value (M$) as a function of the time to maturity (years)

Time	0	1	2	3	4	5	10
W	11,050	11,143	11,216	11,270	11,308	11,332	11,370

Table 8.8 NPV (M$) of a coal plant as opposed to two sets of natural gas stations

Coal-fired plant	1st set gas plants	2nd set gas plants	Total gas plants
11,050.31	1,259.21	3,068.78	4,328.00

8.4 Case 3: An Oil Well

Consider an oil well with known reserves and finite lifetime (for example, owing to expiration of the lease contract). Q_t stands for the rate of extraction at time t. Assume that depletion decreases exponentially at a rate β:

$$Q_t = Q_0 e^{-\beta t}. \tag{8.12}$$

Thus, the well starts producing Q_0 initially but, with $\beta = 0.10$, at the end of the exploitation period 20 years later an amount $0.1353Q_0$ is extracted (this does necessarily imply total exhaustion). Overall, the aggregate extraction over time is given by:

$$\int_0^{20} Q_0 e^{-\beta t} dt = 10Q_0\left(1 - e^{-2}\right) = 8.6466Q_0. \tag{8.13}$$

We are going to compute the value of the project per unit of extraction (i.e. for each barrel depleted). Thus, if just one barrel were to be extracted over the whole useful life of the well, the initial production level would be $Q_0 = 0.11565$ barrels.

We assume that, upon decision to proceed, the facility takes $\tau_1 = 2$ years to build. From then on, the first units of the resource are brought to the surface. Exploitation of the well until date $\tau_2 = 22$ therefore provides a flow of revenues whose present value is:

$$PV = \int_{\tau_1}^{\tau_2} S_t^O Q_0 e^{\beta\tau_1} e^{-(\beta+r)t} dt. \tag{8.14}$$

Note that if production starts at time τ_1 and we want the extraction rate at that time to be Q_0, then the term $Q_0 e^{\beta\tau_1}$ is required inside the integral $Q_0 e^{\beta\tau_1} e^{-\beta\tau_1} = Q_0$. The oil price S_t^O is:

$$S_t^O = \frac{k^O S_m^O}{k^O + \lambda^O}\left(1 - e^{-\left(k^O+\lambda^O\right)t}\right) + S_0^O e^{-\left(k^O+\lambda^O\right)t}. \tag{8.15}$$

Neglecting extraction costs and adopting the values in Table 8.1, hence we compute a net present value:

$$NPV = Q_0 e^{\beta\tau_1}\left[\frac{S_t^O - \frac{k^O S_m^O}{k^O+\lambda^O}}{k^O + \lambda^O + r + \beta}\left[e^{-(k^O+\lambda^O+r+\beta)\tau_1} - e^{-(k^O+\lambda^O+r+\beta)\tau_2}\right]\right.$$
$$\left. + \frac{k^O S_m^O}{(r+\beta)(k^O+\lambda^O)}\left[e^{-(r+\beta)\tau_1} - e^{-(r+\beta)\tau_2}\right]\right]. \tag{8.16}$$

This amounts to using a discount rate $r + \beta$ and multiplying by $Q_0 e^{\beta\tau_1}$. With $\tau_1 = 2$, $\tau_2 = 22$, $\beta = 0.10$, $r = 0.02$, and the data in Table 8.1, we derive $NPV = 78.6809$ \$. If there are costs, they will erode this figure. Consequently, only investments involving a cost lower than 78.68 \$/barrel would be accepted. Note that, according to Table 8.1, the expected oil price approaches 90 \$/barrel. Also, the revenues between $\tau_1 = 2$ and $\tau_2 = 22$ must be discounted at the riskless rate $r = 2$ %.

Table 8.9 Threshold cost to invest in the oil well under different option maturities

Years	0	1	5	10	15	20	25	30
I^*	78.68	74.06	60.48	53.67	51.70	51.00	50.72	50.60

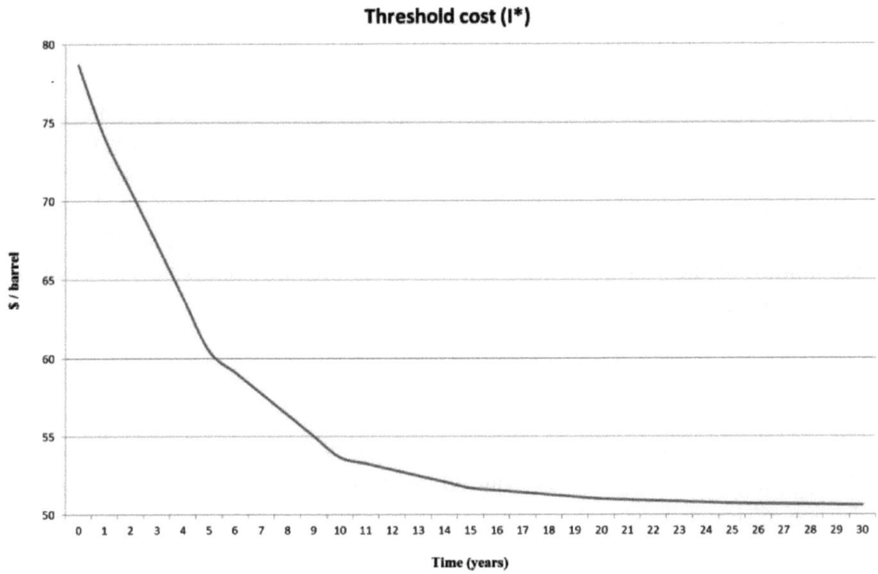

Fig. 8.2 Trigger cost ($/barrel) as a function of the option maturity (years)

Now, assume that we have an American option to invest in this oil well. The maximum investment cost (per barrel extracted) that can be accepted, or the threshold level I^* below which it is optimal to invest, appears in Table 8.9. It depends on the option's time to maturity. The longer the life of the option, the lower the cost must be to trigger investment. Figure 8.2 shows the results.

Chapter 9
Valuation of Energy Assets: Two Risk Factors

9.1 Introduction

This chapter introduces more valuation examples, but now they become more complex in that they account for two risk factors. Thus uncertainty stems from two commodity prices. Specifically, the price of electricity is now governed by a stochastic process akin to those of coal and natural gas. For convenience, we show again the parameter values for each commodity in Table 9.1.

9.2 Case 1: An Advanced Gas/Oil Combined Cycle

Both the price of natural gas and the price of electricity are assumed stochastic. The resulting profit margin for the utility is the so-called spark spread. The data for this case are the same as in the previous chapter. Tables 9.2 and 9.3 show them again for convenience. The gas station has a useful life of 25 years.

Assume that all O&M costs grow at the riskless rate ($r = 0.02$). The present value of all deterministic costs amounts to $I = 765.35$ M\$, and that of the natural gas consumed is 4,646.09 M\$. At the same time, the present value of the electricity produced is 5,764.74 M\$. Hence we get $NPV = 353.31$ M\$. The NPV will fall to zero, however, if the capacity factor drops to 48.62 %; below this level, the NPV would become negative.

Consider the option to invest in this gas-fired plant. Now, both the initial price of electricity, S_t^E, and that of natural gas, S_t^G, depend of the levels reached at the nodes of the lattice. As usual, before computing the value of the option it is necessary to derive the value of the underlying asset. So next we compute the present value of an annuity yielding 1 MWh of electricity over 25 years (from $\tau_1 = t + 3$ to $\tau_2 = t + 28$):

L.M. Abadie and J.M. Chamorro, *Investment in Energy Assets Under Uncertainty*, Lecture Notes in Energy 21, DOI: 10.1007/978-1-4471-5592-8_9,

Table 9.1 Parameter values of the underlying stochastic commodity price processes

Parameter	Electricity ($/MWh)	Natur. gas ($/MWh)	Oil ($/barrel)	Coal ($/tonne)	CO_2 ($/tonne)
$S_0^i - f_i(0)$	65	30	110	90	10
$\frac{k^i S_m^i}{k^i+\lambda^i}$	150	50	90	140	–
$k^i + \lambda^i$	0.06	20	0.30	0.20	−0.05

Table 9.2 Basic parameters of the natural gas-fired power plant

Size (MW)	400	Variable O&M cost ($/MWh)	3.11
Lead time (years)	3	Fixed O&M cost ($/kW)	14.62
Overnight cost ($/kW)	1,003	Heat rate (Btu/kWh)	6,430
Capacity factor (%)	80	Thermal efficiency	0.5307

Table 9.3 Resulting parameters for the natural gas-fired power plant

Yearly output (MWh)	2,803,200	Variable O&M cost (M$/year)	8.718
Heat rate (GJ/MWh)	6.7840	Fixed O&M cost (M$/year)	5.848
Overnight cost (M$)	401.20	CO_2 emissions (tCO_2/year)	1,066,898

$$\begin{aligned} VA_t^E\left(S_t^E\right) &= \frac{S_t^E - \frac{k^E S_m^E}{k^E+\lambda^E}}{k^E + \lambda^E + r}\left[e^{-(k^E+\lambda^E+r)\tau_1} - e^{-(k^E+\lambda^E+r)\tau_2}\right] \\ &+ \frac{k^E S_m^E}{r(k^E + \lambda^E)}\left[e^{-r\tau_1} - e^{-r\tau_2}\right]. \end{aligned} \tag{9.1}$$

The present value of an annuity yielding 1 MWh of natural gas between these dates is:

$$\begin{aligned} VA_t^G\left(S_t^G\right) &= \frac{S_t^G - \frac{k^G S_m^G}{k^G+\lambda^G}}{k^G + \lambda^G + r}\left[e^{-(k^G+\lambda^G+r)\tau_1} - e^{-(k^G+\lambda^G+r)\tau_2}\right] \\ &+ \frac{k^G S_m^G}{r(k^G + \lambda^G)}\left[e^{-r\tau_1} - e^{-r\tau_2}\right]. \end{aligned} \tag{9.2}$$

From these equations, the *NPV* of investing in the gas station at time *t* is:

$$NPV_t\left(S_t^E, S_t^G\right) = 2.8032 VA_t^E\left(S_t^E\right) - \frac{2.8032}{0.5307} VA_t^G\left(S_t^G\right) - I. \tag{9.3}$$

Consider that the time to maturity of the option to invest in the gas plant is 10 years. Taking 24 steps per year the option value amounts to 914.15 M$. This figure is slightly higher than the 888.66 M$ that we derived under a deterministic electricity price. Thus the new source of uncertainty enhances the value of the

Table 9.4 Impact of the capacity factor on the *NPV* and the option value of the gas station

Capacity f. (%)	30	40	50	60	70	80
NPV (M$)	−209.64	−97.05	15.54	128.13	240.72	353.31
Option V.	153.85	284.18	431.45	587.91	749.55	914.15

investment option. Just like in Chap. 8, however, the capacity factor is a major driver of this value. Table 9.4 shows the results.

Table 9.5 shows the impact of the correlation between electricity and gas prices under a given capacity factor (80 %, base case). The *NPV* is independent of ρ_{EG} (= 0.8 in the base case). This does not hold, however, for the option value, which depends inversely on the correlation. In Sect. 8.2, the value of the option was 888.66 M$. Now, with $\rho_{EG} = 1$ it is worth 756.49 M$. This lower option value can be explained as follows: in the model with one risk factor, the electricity price could rise while keeping the same expected value, thus pushing the spark spread upward anyway; in the current model, the rise in the electricity price goes hand in hand with a rise in the gas price (so the seemingly favorable events are no so favorable as before).

The *NPV* and the option value are highly sensitive to changes in the long-term electricity price. The results of this sensitivity analysis are displayed in Table 9.6. A lower average spark spread seriously compromises the profitability of an immediate investment. This in turn dents the value of the option to invest.

We also check how these values are affected by changes in commodity price volatilities (with $\rho_{EG} = 0.80$). For a given gas volatility, wider swings in electricity price leave the *NPV* unchanged but increase the value of the option to invest. When it comes to swings in the input fuel price, given σ_E, a greater value of σ_G depresses the option value. Nonetheless, changes in σ_G have a weaker impact than those in σ_E; the fuel cost is one of several other expenses (unlike electricity, which is the only source of revenue). Note that we do not consider the possibility of flexible plant operation at this time: the plant is assumed to run at 80 % whatever the circumstances in place (Table 9.7).

On the other hand, the effect of volatility changes depends on the correlation between the two prices ρ_{EG}. Table 9.8 addresses this issue; here a correlation $\rho_{EG} = 0.5$ is assumed.

Table 9.5 Impact of the correlation between prices on the *NPV* and the option value

ρ_{EG} (%)	50	60	70	80	90	100
NPV (M$)	353.31	353.31	353.31	353.31	353.31	353.31
Option V.	962.99	946.65	930.21	914.15	899.30	756.49

Table 9.6 Impact of the long-run price of electricity on the *NPV* and the option value

$k^E S^E_m/(k^E + \lambda^E)$	150	140	130	120	110
NPV (M$)	353.31	72.26	−208.77	−489.81	−770.85
Option V.	914.15	648.83	443.57	295.86	194.04

Table 9.7 Impact of price volatilities on the *NPV* and the option value

σ_E	0.15	0.20	0.25	0.30	0.35
NPV (M$)	353.31	353.31	353.31	353.31	353.31
Option V.	874.83	887.66	914.15	954.94	1,006.85
σ_G	0.30	0.35	0.40	0.45	0.50
NPV (M$)	353.31	353.31	353.31	353.31	353.31
Option V.	923.04	917.48	914.15	913.12	914.44

Table 9.8 Impact of price correlation on the option value for different price volatilities

σ_E	0.15	0.20	0.25	0.30	0.35
Option V.	898.60	923.98	962.99	1,013.29	1,071.48
σ_G	0.30	0.35	0.40	0.45	0.50
Option V.	960.13	960.48	962.99	967.65	974.27

Table 9.9 Locus in the prices space over which $NPV = 0$ for different capacity factors

Cap. (%)	S_t^E	50	55	60	65	70	75	80
80	S_t^G	29.66	39.30	48.95	58.58	68.23	77.87	87.51
70	S_t^G	23.33	32.97	42.62	52.26	61.90	71.54	81.19
60	S_t^G	14.90	24.53	34.18	43.82	53.46	63.10	72.75
50	S_t^G	3.09	12.73	22.37	32.01	41.65	51.30	60.94
40	S_t^G	-14.63	-4.99	4.65	14.29	23.94	33.58	43.22

When there is no option to wait, the investment is assessed in terms of the *NPV*. In this case, there is a locus of initial prices in the space (S_t^E, S_t^G) over which we have $NPV = 0$. Table 9.9 displays this locus for different capacity factors. For the *NPV* to remain constant, both the input and the output prices must move in the same direction. The combinations that bring about $NPV = 0$ but include a negative price are not feasible, since prices are bounded from below at zero. Figure 9.1 displays the results.

9.3 Case 2: A New Scrubbed Coal-Fired Station

Now we assess a new coal station under the assumption that coal price and electricity price are both stochastic. For convenience, Tables 9.10 and 9.11 show again the parameter values of the coal plant. It has a useful life of 50 years.

Consider that the plant burns coal with a calorific value of 6,000 kcal/Kg. Then, under 100 % efficiency, each tonne of coal would allow to generate 6.978 MWh of electricity. In other words, under these efficiency circumstances, generating one MWh would require 0.143308 coal tones.

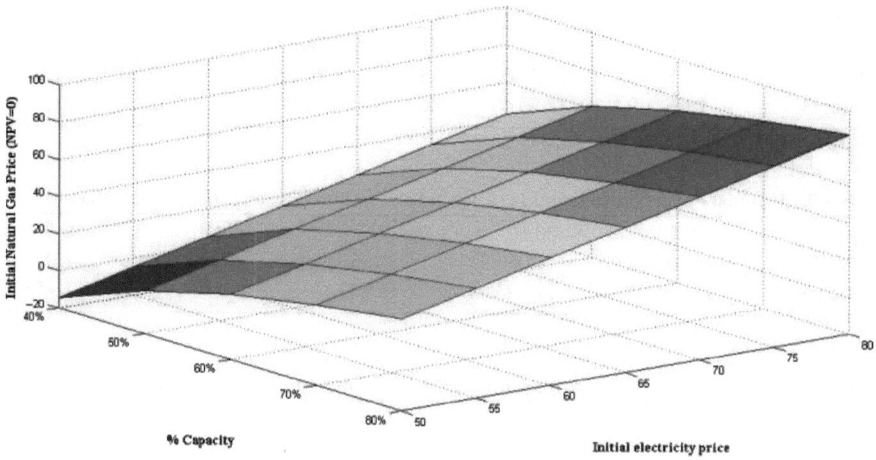

Fig. 9.1 Combinations (S_t^E, S_t^G) for which $NPV = 0$ under different capacity factors

Table 9.10 Basic parameters of the scrubbed coal -fired power plant

Size (MW)	1,200	Variable O&M cost ($/MWh)	4.25
Lead time (years)	4	Fixed O&M cost ($/kW)	29.67
Overnight cost ($/kW)	2,844	Heat rate (Btu/kWh)	8,800
Capacity factor (%)	80	Thermal efficiency	0.3877

Table 9.11 Resulting parameters for the scrubbed coal station

Yearly output (MWh)	8,409,600	Variable O&M cost (M$/year)	35.7408
Heat rate (GJ/MWh)	9.2845	Fixed O&M cost (M$/year)	35.604
Overnight cost (M$)	3,412.80	CO_2 emissions (tCO_2/year)	7,387,086

Assume that all O&M costs grow at the risk-free interest rate ($r = 0.0205$); the present value of all deterministic costs amounts to $I = 6,980.04$ M$. The present value of the coal consumed is 12,404.03 M$, whereas that of the electricity produced reaches 30,434.38 M$. Therefore, we get $NPV = 11,050.31$ M$. A lower capacity factor results in a lower *NPV*; indeed, $NPV = 0$ for an operation level of 25.58 %; below this level, we would get $NPV < 0$.

Regarding the value of the option to invest in this coal station, now the option value is 11, 774.18 M$. This amount is slightly higher than the one we would get in Chap. 8 with a single risk factor, namely 11,370.35 M$. The stochastic character of electricity price enhances the value of the option to invest.

On the other hand, both the *NPV* and the option value change with the capacity factor. Table 9.12 displays the results ($\rho_{EC} = 0.80$). As before, a more intensive utilization of the capital asset translates into higher expected profits now and in the future.

Table 9.12 Impact of the capacity factor on the *NPV* and the option value of the coal plant

Capacity f. (%)	30	40	50	60	70	80
NPV (M$)	898.24	2,928.65	4,959.07	6,989.48	9,019.90	11,050.3
Option V.	1,641.94	3,650.40	5,672.10	7,701.39	9,735.92	11,774.1

Table 9.13 Impact of the correlation between prices on the *NPV* and the option value

ρ_{EC}	10 %	20 %	30 %	40 %	50 %
NPV (M$)	11,050.31	11,050.31	11,050.31	11,050.31	11,050.31
Option V.	11,791.75	11,783.09	11,774.18	11,764.99	11,755.52

Table 9.14 Impact of price volatilities on the *NPV* and the option value of the coal station

σ_E	0.15	0.20	0.25	0.30	0.35
NPV (M$)	11,050.31	11,050.31	11,050.31	11,050.31	11,050.31
Option V.	11,566.39	11,669.32	11,774.18	11,879.44	11,984.29
σ_C	0.10	0.15	0.20	0.25	0.30
NPV (M$)	11,050.31	11,050.31	11,050.31	11,050.31	11,050.31
Option V.	11,781.68	11,777.41	11,774.18	11,771.24	11,771.75

Table 9.15 Price barriers over which $NPV = 0$ for different capacity factors

Cap. (%)	$\hat{S}^E_m$	100	110	120	130	140	150	160
80	$\hat{S}^C_m$	169.81	189.90	209.99	230.08	250.17	270.26	290.35
70	$\hat{S}^C_m$	161.07	181.16	201.25	221.34	241.43	261.52	281.60
60	$\hat{S}^C_m$	149.41	169.50	189.59	209.68	229.77	249.86	269.94
50	$\hat{S}^C_m$	133.08	153.17	173.26	193.35	213.44	233.53	253.62
40	$\hat{S}^C_m$	108.60	128.69	148.78	168.87	188.96	209.04	229.13
30	$\hat{S}^C_m$	67.79	87.88	107.97	128.06	148.15	168.24	188.33
20	$\hat{S}^C_m$	−13.83	6.26	26.35	46.44	66.53	86.62	106.71

Note Both $\hat{S}^E_m$ and $\hat{S}^C_m$ refer to the long-run price levels under risk neutrality

Table 9.13 shows the impact of the correlation between coal and electricity prices (under a capacity factor 80 %). Similarly to the results for the gas-fired station, as the input and output prices are more closely related the *NPV* remains unaffected but the option value decreases (albeit moderately; the cost of fuel as a proportion of the total cost incurred is relatively lower in a coal plant).

The value of the investment opportunity can also be affected by changes in price volatilities. See Table 9.14. Given σ_G, more volatile electricity prices enhance the value of the option to invest. On the contrary, given σ_E, wider swings in coal prices reduce the option value.

When we deal with a now-or-never investment (i.e. there is no delay option) the standard *NPV* rule applies. We can derive the barrier in the space

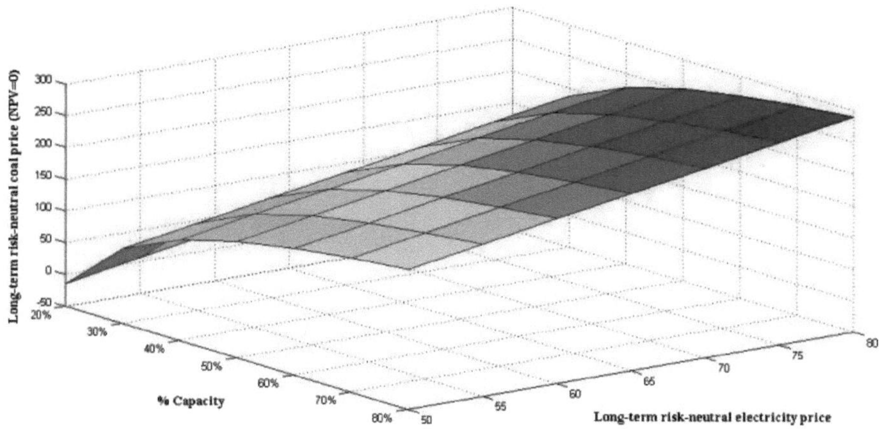

Fig. 9.2 Combinations $(\hat{S}_m^E, \hat{S}_m^C)$ for which $NPV = 0$ under different capacity factors

$(\hat{S}_m^E \equiv k^E S_m^E/(k^E + \lambda^E), \hat{S}_m^C \equiv k^C S_m^C/(k^C + \lambda^C))$ that separates the $NPV > 0$ region from the $NPV < 0$ region; i.e. along this barrier we have $NPV = 0$. The precise combinations of initial prices that set this barrier depend on the capacity factor of the plant. Table 9.15 shows the resulting barriers different capacity factors. As seen in the table, both the input and the output prices must move in the same direction if the *NPV* is to remain the same. Price pairs $(k^E S_m^E/(k^E + \lambda^E), k^C S_m^C/(k^C + \lambda^C))$ resulting in $NPV = 0$ but involving a negative price are not feasible. Figure 9.2 shows the results.

Chapter 10
Valuation of Energy Assets: Three Risk Factors

10.1 Introduction

In this chapter, both the gas plant and the coal station are assumed to operate under carbon constraints. Thus we introduce a new risk factor, namely the price of the carbon emission allowance. Unlike the other two commodity prices, we assume that carbon price follows a standard GBM, which is a non-stationary process. The parameter values for each commodity are shwon once more in Table 10.1.

10.2 Case 1: An Advanced Gas/Oil Combined Cycle

When there is a price on carbon, the owners of gas-fired power plants are naturally interested in the so-called clean spark spread. The data remain the same as before; see Tables 10.2 and 10.3. The gas station has a useful life of 25 years.

According to IPCC (2006), a plant burning natural gas has a carbon emissions factor of 56.1 kg CO_2/GJ. Under 100 % efficiency conditions 3.6 GJ would be consumed per megawatt-hour; hence we get

$$I_G = \frac{0.20196}{E_G} \frac{tCO_2}{\text{MWh}} = 0.3806 \frac{tCO_2}{\text{MWh}}, \tag{10.1}$$

where I_G stands for the emission intensity of the plant (tCO_2/MWh). Therefore, yearly carbon emissions are:

$$2,803,200\,\text{MWh} \times 0.3806 \frac{tCO_2}{\text{MWh}} = 1,066,898\ tCO_2. \tag{10.2}$$

If the carbon price starts initially at $S_0^C = 10$ \$/tCO_2, the present value of the allowances required for emitting 1 tCO_2 per year between τ_1 and τ_2 is:

$$VA_t^A(S_t^A) = \frac{S_0^A}{\alpha^A - \lambda^A - r}\left[e^{(\alpha^A-\lambda^A-r)\tau_2} - e^{(\alpha^A-\lambda^A-r)\tau_1}\right]. \tag{10.3}$$

L.M. Abadie and J.M. Chamorro, *Investment in Energy Assets Under Uncertainty*, Lecture Notes in Energy 21, DOI: 10.1007/978-1-4471-5592-8_10,

Table 10.1 Parameter values of the underlying stochastic commodity price processes

Parameter	Electricity ($/MWh)	Natural gas ($/MWh)	Oil ($/barrel)	Coal ($/t)	CO_2 ($/t)
$S_0^i - f_i(0)$	65	30	110	90	10
$\frac{k^i S_m^i}{k^i+\lambda^i}$	150	50	90	140	–
$k^i + \lambda^i$	0.06	20	0.30	0.20	−0.05

Table 10.2 Basic parameters of the natural gas-fired power plant

Size (MW)	400	Variable O&M cost ($/MWh)	3.11
Lead time (years)	3	Fixed O&M cost ($/kW)	14.62
Overnight cost ($/kW)	1,003	Heat rate (Btu/kWh)	6,430
Capacity factor (%)	80	Thermal efficiency	0.5307

Table 10.3 Resulting parameters for the natural gas-fired power plant

Yearly output (MWh)	2,803,200	Variable O&M cost (M$/year)	8.718
Heat rate (GJ/MWh)	6.7840	Fixed O&M cost (M$/year)	5.848
Overnight cost (M$)	401.20	CO_2 emissions (tCO_2/year)	1,066,898

With $\alpha^A - \lambda^A = 0.05$, $\tau_1 = t + 3$, $\tau_2 = t + 28$, and $r = 0.02$, we get $VA_t^A(S_t^A) = 407.40$ \$. Therefore, the present value of the allowances for total emissions is 434.65 M$.

If O&M costs grow at the riskless rate, r, the present value of all deterministic costs amounts to $I = 765.35$ M\$. Natural gas expenses have a present value 4,646.09 M\$. And the present value of the electricity produced is 5,764.74 M\$. Hence we compute $NPV = -81.35$ M\$. Now, to get $NPV = 0$ requires a capacity factor of 93.96 %; below this (extremely high) level, the NPV would be negative.

Consider the option to invest in this gas-fired plant. Now, the initial prices of the three commodities (S_t^E, S_t^G, and S_t^A) depend on the levels reached at the nodes of the three-dimensional lattice. As usual, before computing the value of the option it is necessary to derive the value of the underlying asset.

The present value of an annuity yielding 1 MWh of electricity over 25 years (from $\tau_1 = 3$ to $\tau_2 = 28$) is:

$$VA_t^E\left(S_t^E\right) = \frac{S_t^E - \frac{k^E S_m^E}{k^E+\lambda^E}}{k^E + \lambda^E + r}\left[e^{-(k^E+\lambda^E+r)\tau_1} - e^{-(k^E+\lambda^E+r)\tau_2}\right] + \frac{k^E S_m^E}{r(k^E + \lambda^E)}\left[e^{-r\tau_1} - e^{-r\tau_2}\right]. \tag{10.4}$$

The present value of an annuity yielding 1 MWh of natural gas between these dates is:

$$VA_t^G\left(S_t^G\right) = \frac{S_t^G - \frac{k^G S_m^G}{k^G+\lambda^G}}{k^G + \lambda^G + r}\left[e^{-(k^G+\lambda^G+r)\tau_1} - e^{-(k^G+\lambda^G+r)\tau_2}\right] + \frac{k^G S_m^G}{r(k^G + \lambda^G)}\left[e^{-r\tau_1} - e^{-r\tau_2}\right]. \quad (10.5)$$

The NPV of investing in the gas-fired station at time t is:

$$NPV_t\left(S_t^E, S_t^G, S_t^A\right) = 2.8032\, VA_t^E\left(S_t^E\right) - \frac{2.8032}{0.5307} VA_t^G\left(S_t^G\right) - \frac{2.8032 \times 0.20196}{0.5307} VA_t^A\left(S_t^A\right) - I. \quad (10.6)$$

If the option's time to expiration is 10 years and we take 6 steps per year, we get an option value of 534.09 M\$. This is far less than the 914.15 M\$ in Chap. 9. Thus the new environmental restriction has a significant impact on the investment opportunity. Indeed, we have already seen that its net present value is negative: $NPV = -81.35$ M\$.

The capacity factor continues to be a key parameter; see Table 10.4.

Table 10.5 displays the sensitivity of the NPV and the option value to the long-term electricity price. A lower average spark spread seriously compromises the profitability of an immediate investment. This in turn dents the value of the option to invest.

We also check how changes in commodity price volatilities affect option value. See Table 10.6. For given σ_G and σ_A, wider swings in electricity price increase the value of the investment option. However, given σ_E and σ_A, a greater value of σ_G decreases the option value. Given σ_E and σ_G, more volatile carbon prices enhance the value of the investment option. Note that $\rho_{EG} = 0.80$ and $\rho_{EA} = 0.40$ in the base case; carbon cost is less important than gas cost, and it is less correlated with electricity price.

Table 10.4 Impact of the capacity factor on the NPV and the option value of the gas station

Capacity factor (%)	75	80	85	90
NPV (M\$)	−110.47	−81.35	−52.22	−23.09
Option value	482.46	534.09	586.17	638.74

Table 10.5 Impact of the long-run price of electricity on the NPV and the option value

S_m^E	130	140	150	160	170
NPV (M\$)	−643.42	−362.39	−81.35	199.69	480.73
Option value	225.65	352.25	534.09	771.77	1,045.96

Table 10.6 Impact of price volatilities on the NPV and the option value

σ_E	0.15	0.20	0.25	0.30	0.35
NPV (M$)	−81.35	−81.35	−81.35	−81.35	−81.35
Option value	444.64	481.70	534.09	599.72	674.79
σ_G	0.30	0.35	0.40	0.45	0.50
NPV (M$)	−81.35	−81.35	−81.35	−81.35	−81.35
Option value	552.99	542.18	534.09	528.59	526.03
σ_A	0.30	0.35	0.40	0.45	0.50
NPV (M$)	−81.35	−81.35	−81.35	−81.35	−81.35
Option value	509.83	520.28	534.09	549.78	567.21

Table 10.7 Impact of price correlation on the option value for different price volatilities

σ_E	0.15	0.20	0.25	0.30	0.35
Option value	574.96	626.42	690.03	761.83	838.65
σ_G	0.30	0.35	0.40	0.45	0.50
Option value	671.15	680.17	690.03	700.23	710.60
σ_A	0.30	0.35	0.40	0.45	0.50
Option value	663.34	675.13	690.03	707.41	726.54

The impact of volatility is very limited due to the effect of the correlation between gas and electricity prices, ρ_{EG}. Table 10.7 assumes a correlation $\rho_{EG} = 0$.

Table 10.8 shows the value of the option as a function of fuel and electricity price volatilities for a given level of carbon price volatility ($\sigma_A = 0.40$). Figure 10.1 displays the results.

Even if volatility were zero the option to wait can be valuable since there is an optimal time to invest in a deterministic framework. The clean spark spread can well be negative in the first years, but time is on its side; after a few years, it turns to positive. Table 10.9 shows the value of the option as a function of σ_E and σ_G for $\sigma_A = 0.20$. Figure 10.2 displays the results.

Table 10.8 Impact of fuel and electricity price volatilities on the option value ($\sigma_A = 0.40$)

	σ_E						
σ_G	0.05	0.10	0.15	0.20	0.25	0.30	0.35
0.20	310.24	426.71	467.17	517.74	585.10	662.73	745.98
0.25	290.20	416.67	457.15	503.68	566.92	641.83	723.53
0.30	276.93	410.62	450.20	493.66	552.99	624.92	704.59
0.35	269.41	407.52	445.94	486.48	542.18	611.12	688.48
0.40	266.48	407.13	444.64	481.70	534.09	599.72	674.79
0.45	266.95	408.96	445.76	479.76	528.59	591.41	663.13
0.50	269.90	412.80	449.19	480.84	526.03	584.63	654.39

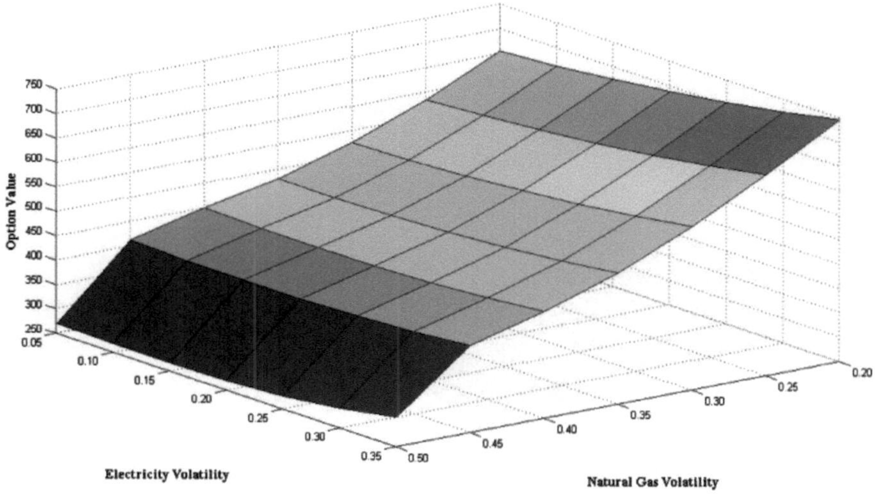

Fig. 10.1 Impact of fuel and electricity price volatilities on the option value ($\sigma_A = 0.40$)

Table 10.9 Impact of electricity and gas price volatilities on the option value ($\sigma_A = 0.20$)

	σ_E						
σ_G	0.05	0.10	0.15	0.20	0.25	0.30	0.35
0.20	235.38	337.90	394.72	470.56	557.32	648.60	741.34
0.25	219.41	330.74	383.80	454.05	537.37	626.92	718.83
0.30	210.52	326.50	377.70	442.17	521.63	608.74	699.16
0.35	207.20	325.44	374.37	434.68	509.44	593.47	681.85
0.40	207.94	327.32	373.88	429.74	500.22	580.79	667.02
0.45	211.53	331.69	376.44	428.08	494.84	572.03	654.60
0.50	216.91	338.21	381.41	430.02	492.79	565.14	645.55

If it is not possible to delay the investment, the latter must be assessed in terms of its NPV. With three risk factors, there is a surface of initial prices in the space (S_0^E, S_0^G, S_0^A) over which we have $NPV = 0$. The upper part of Table 10.10 displays this locus for different capacity factors under $S_0^A = 10$ \$/tCO_2. The lower part of the table, instead, shows the pairs $(\hat{S}_m^E \equiv k^E S_m^E/(k^E + \lambda^E), S_0^G)$ yielding $NPV = 0$. Combinations involving a negative price are not feasible.

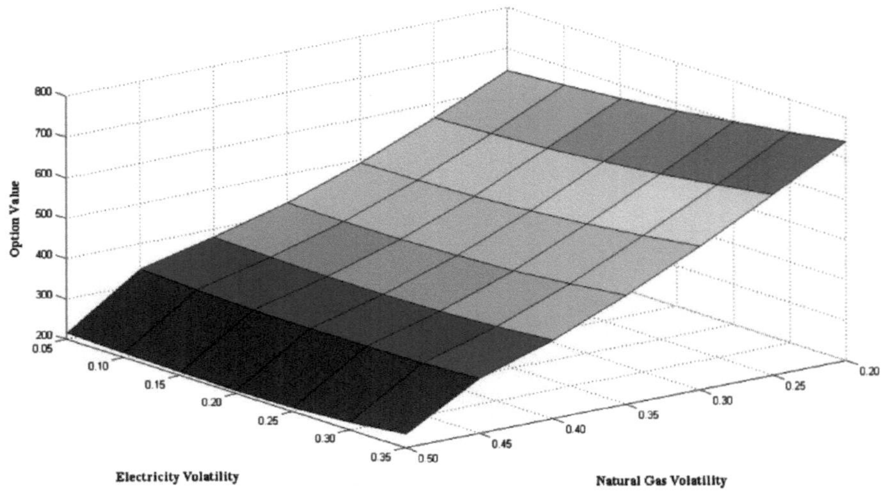

Fig. 10.2 Impact of electricity and gas price volatilities on the option value ($\sigma_A = 0.20$)

Table 10.10 Locus in the prices space over which NPV = 0 for different capacity factors

Capacity factors	S_0^E	50	55	60	65	70	75	80
80 %	S_0^G	−5.51	4.13	13.78	23.42	33.06	42.70	52.34
70 %	S_0^G	−11.84	−2.19	7.45	17.09	26.73	36.37	46.02
Capacity factors	$\hat{S}_m^E$	140	150	160	170	180	190	200
80 %	S_0^G	0.68	23.42	46.16	68.90	91.64	114.38	137.12
70 %	S_0^G	−5.65	17.09	39.83	62.57	85.31	108.05	130.79

10.3 Case 2: A New Scrubbed Coal-Fired Station

Here a new coal station is evaluated assuming that the prices of coal, electricity, and carbon are all stochastic. For convenience, Tables 10.11 and 10.12 show again the parameter values of the coal plant. It has a useful life of 50 years.

Consider that the plant burns coal with a calorific value of 6,000 kcal/kg. Under 100 % efficiency, each tonne of coal would allow to generate 6.978 MWh of electricity. In other words, under these efficiency circumstances, generating one MWh would require 0.143308 coal tones.

According to IPCC (2006), a plant burning bituminous coal has a carbon emissions factor of 94.6 kg CO_2/GJ. Under 100 % efficiency conditions 3.6 GJ would be consumed per megawatt-hour; hence we get

$$I_C = \frac{0.34056}{E_C}\frac{tCO_2}{\text{MWh}} = 0.878\frac{tCO_2}{\text{MWh}}. \tag{10.7}$$

Table 10.11 Basic parameters of the scrubbed coal-fired power plant

Size (MW)	1,200	Variable O&M cost ($/MWh)	4.25
Lead time (years)	4	Fixed O&M cost ($/kW)	29.67
Overnight cost ($/kW)	2,844	Heat rate (Btu/kWh)	8,800
Capacity factor (%)	80	Thermal efficiency	0.3877

Table 10.12 Resulting parameters for the scrubbed coal station

Yearly output (MWh)	8,409,600	Variable O&M cost (M$/year)	35.7408
Heat rate (GJ/MWh)	9.2845	Fixed O&M cost (M$/year)	35.604
Overnight cost (M$)	3,412.80	CO_2 emissions (tCO_2/year)	7,387,086

Thus, yearly carbon emissions are:

$$8,409,600\,\text{MWh} \times 0.878\frac{tCO_2}{\text{MWh}} = 7,387,086\,tCO_2. \tag{10.8}$$

If the carbon price starts initially at $S_0^C = 10$ $/$tCO_2$, the present value of the allowances required for emitting 1 tCO_2 per year between τ_1 and τ_2 is:

$$VA_t^A\left(S_t^A\right) = \frac{S_0^A}{\alpha^A - \lambda^A - r}\left[e^{(\alpha^A - \lambda^A - r)\tau_2} - e^{(\alpha^A - \lambda^A - r)\tau_1}\right]. \tag{10.9}$$

With $\alpha^A - \lambda^A = 0.05$, $\tau_1 = t + 3$, $\tau_2 = t + 53$, and $r = 0.02$, we get $VA_t^A\left(S_t^A\right) = 1,308.53$ $. Therefore, the present value of the allowances for total emissions is 9,661.71 M$.

Assume that O&M costs grow at the risk-free interest rate $(r = 0.02)$; the present value of all deterministic costs amounts to $I =$ 6,980.04 M$, while that of the coal consumed is 12,404.03 M$. Revenues from electricity sales reach 30,434.38 M$. Therefore, we get $NPV =$ 1,388.60 M$. If the operation level were 63.12 % we would get $NPV = 0$; for lower capacity factors the NPV would be negative.

Now the value of the option to invest in this coal station is 4,840.60 M$. Note that there was no carbon cost in Chap. 9, which resulted in an option value of 11,774.18 M$.

Table 10.13 displays the changes in the NPV and the option value under different capacity factors. Again, more operation hours enhance current and prospective profits. Yet these values are much lower than in the absence of a carbon price.

Table 10.13 Impact of the capacity factor on the NPV and the option value of the coal plant

Capacity factor (%)	30	40	50	60	70	80
NPV (M$)	−2,724.9	−1,902.2	−1,079.5	−256.80	565.90	1,388.60
Option value	297.19	1,014.72	1,906.68	2,858.78	3,841.21	4,840.60

Table 10.14 Impact of the long-run price of electricity on the NPV and the option value

$k^E S^E_m/(k^E + \lambda^E)$	130	140	150	160	170
NPV (M$)	−2,019.9	−315.64	1,388.60	3,092.8	4,797.1
Option value	2,588.62	3,662.83	4,840.60	6,100.29	7,425.57

The NPV and the option value are very sensitive to the long-term electricity price $\hat{S}^E_m \equiv k^E S^E_m/(k^E + \lambda^E)$; see Table 10.14. A higher clean spark spread raises the profitability of an immediate investment and enhances the value of the option to invest.

The option value is also affected by changes in commodity price volatilities. See Table 10.15. All else constant, a higher σ_E tends to increase the value of the option. Conversely, higher fuel cost volatility σ_C typically decreases it. And more volatile carbon prices enhance the value of the investment option. Note that coal and electricity price display mean reversion, while carbon price follows a non-stationary process.

Anyway the impact of coal volatility σ_C is rather limited because of the effect of the correlation with electricity prices, ρ_{EC}. For example, assuming $\rho_{EC} = 0$ we derive the values in Table 10.16.

Instead, the value of the option to invest in the coal plant is strongly affected by carbon price volatility, σ_A. Table 10.17 shows the option value as a function of the other two volatilities under the cross correlations in the base case and $\sigma_A = 0.40$. Figure 10.3 displays the results.

Table 10.15 Impact of price volatilities on the NPV and the option value

σ_E	0.15	0.20	0.25	0.30	0.35
NPV (M$)	1,388.60	1,388.60	1,388.60	1,388.60	1,388.60
Option value	4,841.11	4,828.92	4,840.60	4,875.77	4,936.68
σ_C	0.10	0.15	0.20	0.25	0.30
NPV (M$)	1,388.60	1,388.60	1,388.60	1,388.60	1,388.60
Option value	4,854.27	4,843.15	4,840.60	4,839.51	4,839.61
σ_A	0.30	0.35	0.40	0.45	0.50
NPV (M$)	1,388.60	1,388.60	1,388.60	1,388.60	1,388.60
Option value	3,644.14	4,242.92	4,840.60	5,424.78	5,988.28

Table 10.16 Impact of price correlation on the option value for different price volatilities

σ_E	0.15	0.20	0.25	0.30	0.35
Option value	4,850.72	4,842.91	4,856.75	4,892.94	4,954.49
σ_C	0.10	0.15	0.20	0.25	0.30
Option value	4,836.92	4,852.26	4,856.75	4,860.72	4,864.70
σ_A	0.30	0.35	0.40	0.45	0.50
Option value	3,664.09	4,260.36	4,856.75	5,439.55	6,002.05

Table 10.17 Impact of fuel and electricity price volatilities on the option value ($\sigma_A = 0.40$)

	σ_E						
σ_C	0.05	0.10	0.15	0.20	0.25	0.30	0.35
0.05	4,203.4	4,754.0	4,983.8	5,123.6	5,241.7	5,360.1	5,488.7
0.10	4,692.2	4,874.5	4,843.8	4,835.9	4,854.2	4,898.3	4,972.2
0.15	4,681.1	4,874.0	4,841.5	4,830.2	4,843.1	4,879.2	4,941.6
0.20	4,594.1	4,872.6	4,841.1	4,828.9	4,840.6	4,875.7	4,936.6
0.25	4,513.7	4,871.9	4,842.0	4,828.9	4,839.5	4,874.0	4,933.9
0.30	4,450.8	4,871.9	4,844.2	4,830.1	4,839.6	4,873.4	4,932.4
0.35	4,401.2	4,872.0	4,847.4	4,832.6	4,840.9	4,874.1	4,932.2

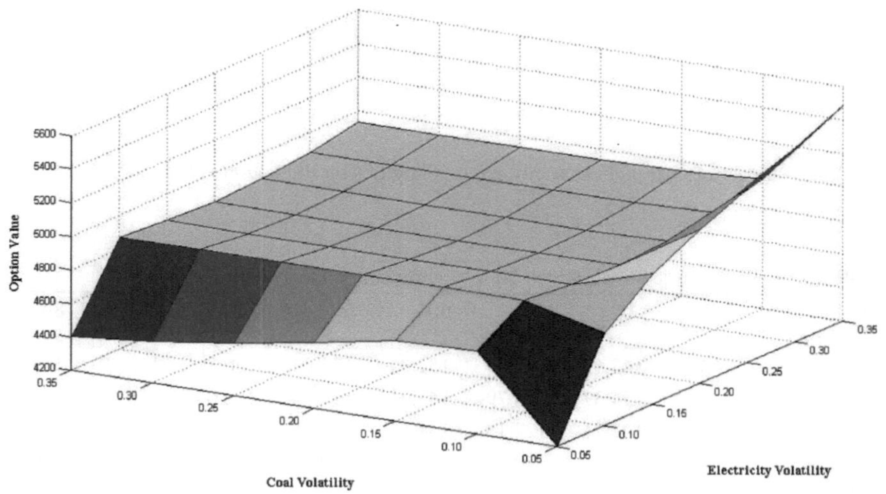

Fig. 10.3 Impact of coal and electricity price volatilities on the option value ($\sigma_A = 0.40$)

Table 10.18 displays the option value as a function of the two other volatilities under $\sigma_A = 0.20$. Note that even under zero volatility the investment option can be valuable since in a deterministic framework it may be optimal to wait. Figure 10.4 shows the results.

Table 10.18 Impact of fuel and electricity price volatilities on the option value ($\sigma_A = 0.20$)

	σ_E						
σ_C	0.05	0.10	0.15	0.20	0.25	0.30	0.35
0.05	2,107.6	2,479.2	2,687.5	2,853.3	3,020.1	3,202.0	3,397.4
0.10	2,393.2	2,396.9	2,392.7	2,448.4	2,552.2	2,694.6	2,868.4
0.15	2,423.6	2,396.6	2,385.5	2,425.4	2,507.9	2,630.8	2,787.4
0.20	2,393.8	2,398.0	2,385.0	2,421.9	2,501.4	2,621.4	2,774.9
0.25	2,360.4	2,400.8	2,386.1	2,421.0	2,498.6	2,617.3	2,769.5
0.30	2,333.5	2,405.1	2,388.9	2,421.9	2,497.7	2,615.6	2,766.3
0.35	2,312.7	2,411.2	2,393.4	2,424.5	2,498.7	2,615.5	2,765.0

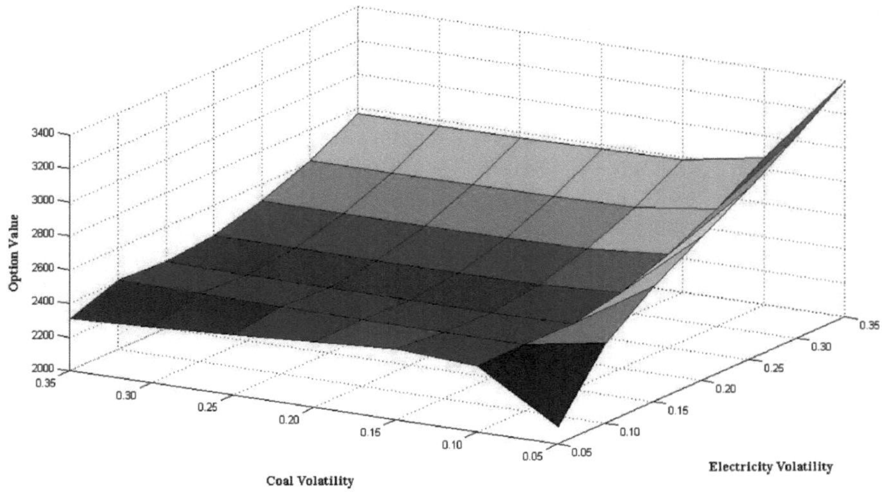

Fig. 10.4 Impact of coal and electricity price volatilities on the option value ($\sigma_A = 0.20$)

Table 10.19 Price barriers over which NPV = 0 for different capacity factors

Capacity factors	S_0^E	50	55	60	65	70	75	80
80 %	S_0^C	143.12	147.54	151.95	156.37	160.79	165.20	169.62
70 %	S_0^C	134.37	138.79	143.21	147.62	152.04	156.46	160.87
60 %	S_0^C	122.71	127.13	131.55	135.96	140.38	144.80	149.21
Capacity factors	$\hat{S}_m^E$	110	120	130	140	150	160	170
80 %	$\hat{S}_m^C$	76.01	96.10	116.19	136.28	156.37	176.46	196.55
70 %	$\hat{S}_m^C$	67.27	87.36	107.44	127.53	147.62	167.71	187.80
60 %	$\hat{S}_m^C$	55.61	75.70	95.78	115.87	135.96	156.05	176.14

Note Both $\hat{S}_m^E$ and $\hat{S}_m^C$ refer to the long-run price levels under risk neutrality

When there is no option to delay investment the standard NPV rule applies. In this case, there is a collection of initial commodity prices, S_0^E, S_o^C, and S_0^A, for which we exactly get $NPV = 0$. A subset of them appears in the upper part of Table 10.19 (under $S_0^A = 10$ \$/tCO_2 and different capacity factors). For the NPV to stay the same, both the fuel and the electricity prices must move in the same direction. The lower part of the table, instead, shows the pairs ($\hat{S}_m^E \equiv k^E S_m^E/(k^E + \lambda^E)$, $\hat{S}_m^C \equiv k^C S_m^C/(k^C + \lambda^C)$) that yield $NPV = 0$.

Chapter 11
Value Maximization and Optimal Management of Energy Assets

11.1 Introduction

There can be situations in which current decisions on how to operate a facility will have no impact on future decisions. Consider, for example, an industrial boiler which can run alternatively on coal and natural gas. Further, assume that there are no switching costs between modes of operation. Thus, the manager can find it optimal today to burn coal (if this is the most profitable choice), and this course of action would not affect the decision to be made tomorrow. Note anyway the assumption that there is a dual fuel boiler in the first place. This means that flexibility is important not only at the operation stage but at previous stages as well (such as the concept or design stages); see de Neufville and Scholtes (2011).

In other cases current decisions affect future ones, and this must be taken into account in the decision making process. For instance, assume that switching costs are not negligible. If so, leaning on the mode that allows a paltry saving may not be in our best interest if there is a high probability that we will have to reverse course in the near future and the costs to getting back to the original state more than offset the initial saving. Another, more extreme, case can arise when management decides to close down a unit (e.g. a facility, or a mine) temporarily and there are sizeable costs to opening it up again. In this case, it will be optimal to keep it open despite the losses (at least for a while) if the chance of getting profits in the future is big.

A huge number of different typologies fall within this framework. Yet another example is that of an ore mine (or oil well) in which extraction costs rise as the stock is depleted. This feature will have to be accounted for when setting the optimal extraction path alongside uncertainty in commodity prices.

L.M. Abadie and J.M. Chamorro, *Investment in Energy Assets Under Uncertainty*,
Lecture Notes in Energy 21, DOI: 10.1007/978-1-4471-5592-8_11,

11.2 Case 1: A Natural Gas-Fired Power Plant ("On" or "Off"; no Switching Costs)

We develop the same valuation approach as in Chap. 10, but now the gas station can switch between mode "on" (if the clean spark spread is positive) and mode "off" (when the spread is negative). For simplicity, at this time there are no switching costs between states or modes. Tseng and Lin (2007) have addressed this issue; they conclude that their impact on valuation is not huge.

Next we generate correlated random variables for electricity, natural gas, and emission allowance prices. Regarding electricity we have:

$$\begin{aligned}\left(S^E_{t+\Delta t} - S^E_t\right) + \left(f_E(t+\Delta t) - f_E(t)\right) = {} & \left[k^E S^E_m - k^E\left(S^E_t - f_E(t)\right)\right]\Delta t \\ & + \sigma_E\left(S^E_t - f_E(t)\right)\sqrt{\Delta t}\,\in^E_t .\end{aligned} \tag{11.1}$$

Using the deseasonalised series:

$$D^E_t \equiv S^E_t - f_E(t), \tag{11.2}$$

for the electricity price we get:

$$D^E_{t+\Delta t} - D^E_t = \left(k^E S^E_m - k^E D^E_t\right)\Delta t + \sigma_E D^E_t \sqrt{\Delta t}\,\in^1_t . \tag{11.3}$$

For (deseasonalised) natural gas price, the random path follows:

$$S^G_{t+\Delta t} - S^G_t = \left(k^G S^G_m - k^G S^G_t\right)\Delta t + \sigma_G D^G_t \sqrt{\Delta t}\left[\in^1_t \rho_{EG} + \in^2_t \sqrt{1-\rho^2_{EG}}\right]. \tag{11.4}$$

For the emission allowance price we compute:

$$\begin{aligned} lnS^A_{t+\Delta t} = {} & lnS^A_t + \left(\alpha - \frac{\sigma^2_A}{2}\right)\Delta t \\ & + \sigma_A\sqrt{\Delta t}\left[\in^1_t \rho_{EA} + \in^2_t \frac{\rho_{GA} - \rho_{EA}\rho_{EG}}{\sqrt{1-\rho^2_{EG}}} + \in^3_t \sqrt{1-\rho^2_{EA} - \frac{\left(\rho_{GA}-\rho_{EA}\rho_{EG}\right)^2}{1-\rho^2_{EG}}}\right].\end{aligned} \tag{11.5}$$

We consider an investment in a gas station with installed capacity 400 MW, $I_G = 0.3806$, $E_G = 0.5307$, and capacity factor 80 %. First we are going to derive its value by Monte Carlo simulation under the assumption that it operates round the clock. In this case, with 60 time steps per year (i.e. $\Delta t = 1/60$), the plant produces 2,803,200 MWh in a year, and 46,720 MWh in each step. Therefore, the present value of the profit margins is given by:

$$\frac{1}{N}\sum_{i=1}^{N}\sum_{j=\frac{3}{\Delta t}}^{j=\frac{28}{\Delta t}} 46,720\left[\left(S^E(i,j) - \frac{S^G(i,j)}{0.5307} - \frac{0.20196 S^A(i,j)}{0.5307}\right)e^{-rj\Delta t} - cv^G\right]. \tag{11.6}$$

Here $N = 20,000$ denotes the number of simulation runs, and $cv^G = 3.11$ stands for the operation variable cost of the natural gas plant. Equation (11.6) assumes a lead time of 3 years; henceforth the (base load) plant has a useful life of 25 years.

Now, if the plant can be switched either "on" or "off" (depending on the clean spark spread), then management holds an option (to produce); obviously this option can be exercised or not. The value of the plant can then be computed as:

$$\frac{1}{N}\sum_{i=1}^{N}\sum_{j=\frac{3}{\Delta t}}^{j=\frac{28}{\Delta t}} 46,720\, max\left[\left(S^E(i,j) - \frac{S^G(i,j)}{0.5307} - \frac{0.20196 S^A(i,j)}{0.5307} e^{-rj\Delta t} - cv^G\right), 0\right]. \tag{11.7}$$

From this (gross) present value we must deduce that of construction costs and fixed costs, which amount to $I = 547.40$ M$.

We run 20,000 simulations. We check the cross correlations between commodity prices in the random samples of the last time step. Table 11.1 compares the theoretical values with the simulated ones. They match each other to a high degree. As could be expected, Monte Carlo simulation provides better results when the underlying stochastic processes show mean reversion.

Concerning the valuation of the plant, Table 11.2 displays the results in both scenarios, i.e. without and with operation flexibility. The flexible operation brings about a positive present value, 594.26 M$. This is accomplished by ceasing operations whenever the clean spark spread becomes negative. The value of flexibility amounts to 594.26 – (– 81.35) = 675.61 M$.

Next we assess how the value of flexibility is affected by the allowance price volatility. Table 11.3 displays the results. As emission allowance prices become more volatile the value of the plant (under flexible operation) grows higher.

Now we consider the case in which the three base volatilities are scaled down in the same proportion. According to Table 11.4, decreasing volatilities leads to

Table 11.1 Theoretical and simulated correlations between commodity prices

Correlation coefficient	Theoretical	Simulated
ρ_{EG}	0.80	0.8007
ρ_{EA}	0.40	0.4058
ρ_{GA}	0.25	0.2601

Table 11.2 Present value of the gas plant under flexible and rigid operation

	Rigid operation	Flexible operation
Cumulative spreads	466.05	1,141.66
Fixed O&M costs	−146.20	−146.20
Overnight cost	−401.20	−401.20
Present value (M$)	−81.35	594.26

Table 11.3 Sensitivity of the plant value to allowance price volatility

σ_A	0.20	0.30	0.40	0.50	0.60
Present value (M$)	554.24	566.47	594.26	631.07	670.07

Table 11.4 Sensitivity of the plant value to volatility

σ_E, σ_G, σ_A	100 %	75 %	50 %	25 %	10 %
Present value (M$)	594.26	368.22	159.87	−9.70	−63.82

lower plant values; see also Fig. 11.1. Indeed, the present value of the plant approaches the analytical solution for the base load plant, $NPV = -81.35$ M$. In the limit, under null volatilities, this *NPV* would be reached and the investment would not be undertaken. Investing becomes profitable when volatility levels surpass around 25 % of those in the base case.

11.3 Case 2: A Coal-Fired Power Plant ("On" or "Off"; no Switching Costs)

The valuation approach in Chap. 10 is followed again here. The key difference, however, is that the coal station can switch between mode "on" (if the clean dark spread is positive) and mode "off" (when the spread is negative).

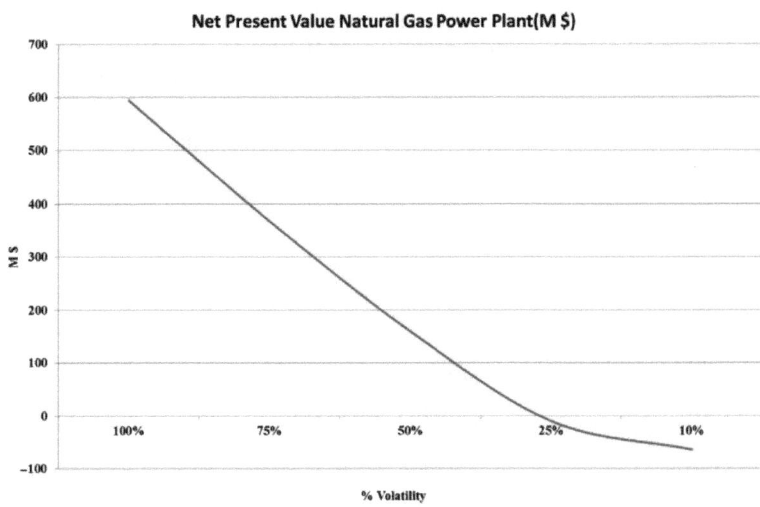

Fig. 11.1 *NPV* of the gas plant as a function of commodity volatilities

Correlated random variables for electricity, natural gas, and emission allowance prices are generated following a similar approach to that in Sect. 11.1. Regarding electricity we have:

$$\begin{aligned}\left(S^E_{t+\Delta t} - S^E_t\right) + \left(f_E(t+\Delta t) - f_E(t)\right) = {} & \left[k^E S^E_m - k^E\left(S^E_t - f_E(t)\right)\right]\Delta t \\ & + \sigma_E\left(S^E_t - f_E(t)\right)\sqrt{\Delta t}\,\epsilon^E_t\,.\end{aligned} \tag{11.8}$$

Using the deseasonalised series:

$$D^E_t \equiv S^E_t - f_E(t), \tag{11.9}$$

for the electricity price we have:

$$D^E_{t+\Delta t} - D^E_t = \left(k^E S^E_m - k^E D^E_t\right)\Delta t + \sigma_E D^E_t \sqrt{\Delta t}\,\epsilon^1_t\,. \tag{11.10}$$

For (deseasonalised) coal price, the random path follows:

$$S^C_{t+\Delta t} - S^C_t = \left(k^C S^C_m - k^C S^C_t\right)\Delta t + \sigma_C D^C_t \sqrt{\Delta t}\left[\epsilon^1_t\,\rho_{EC} + \epsilon^2_t\,\sqrt{1-\rho^2_{EC}}\right]. \tag{11.11}$$

For the emission allowance price we compute:

$$\begin{aligned} lnS^A_{t+\Delta t} = {} & lnS^A_t - \left(\alpha - \frac{\sigma^2_A}{2}\right)\Delta t \\ & + \sigma_A\sqrt{\Delta t}\left[\epsilon^1_t\,\rho_{EA} + \epsilon^2_t\,\frac{\rho_{CA} - \rho_{EA}\rho_{EC}}{\sqrt{1-\rho^2_{EC}}} + \epsilon^3_t\,\sqrt{1 - \rho^2_{EA} - \frac{\left(\rho_{CA} - \rho_{EA}\rho_{EC}\right)^2}{1-\rho^2_{EC}}}\right].\end{aligned} \tag{11.12}$$

We consider an investment in a coal station with installed capacity 1,200 MW, $I_C = 0.8784$, $E_C = 0.3877$, and capacity factor 80 %. First we compute its value by simulation assuming that it operates round the clock. With 30 time steps per year (i.e. $\Delta t = 1/30$), the plant produces 8,409,600 MWh in a year, and 280,320 MWh in each step. Therefore, the present value of the profit margins is given by:

$$\frac{1}{N}\sum_{i=1}^{N}\sum_{j=\frac{4}{\Delta t}}^{j=\frac{54}{\Delta t}} 280,320\left[\left(S^E(i,j) - \frac{S^C(i,j)}{0.3877\times 6.978} - \frac{0.34056 S^A(i,j)}{0.3877}\right)e^{-rj\Delta t} - cv^C\right]. \tag{11.13}$$

Here $N = 20,000$ denotes the number of simulation runs, and $cv^C = 4.25$ stands for the operation variable cost of the coal plant. Equation (11.13) assumes a lead time of 4 years; henceforth the (base load) plant has a useful life of 50 years.

Now, if the plant can be switched either "on" or "off" (depending on the clean dark spread), then management holds an option (to produce); obviously this option can be exercised or not. The value of the plant can then be computed as:

Table 11.5 Present value of the coal plant under flexible and rigid operation

	Rigid operation	Flexible operation
Cumulative spreads	6,581.60	12,999.37
Fixed O&M costs	−1,780.20	−1,780.20
Overnight cost	−3,412.80	−3,412.80
Present value (M$)	1,388.60	7,806.37

Table 11.6 Sensitivity of the plant value to allowance price volatility

σ_A	0.20	0.30	0.40	0.50	0.60
Present value (M$)	5,664.15	6,724.41	7,806.37	8,735.19	9,464.08

Table 11.7 Sensitivity of the plant value to volatility

σ_E, σ_G, σ_A	100 %	75 %	50 %	25 %	10 %
Present value (M$)	7,806.37	6,173.37	4,439.49	2,819.23	2,097.76

$$\frac{1}{N}\sum_{i=1}^{N}\sum_{j=\frac{4}{\Delta t}}^{j=\frac{54}{\Delta t}} 280,320 max\left[\left[\left(S^E(i,j) - \frac{S^C(i,j)}{0.3877\times 6.978} - \frac{0.34056S^A(i,j)}{0.3877}\right)e^{-rj\Delta t} - cv^C\right],0\right]. \tag{11.14}$$

From this (gross) present value we must deduce that of construction costs and fixed costs, which amount to $I = 5,193.00$ M$.

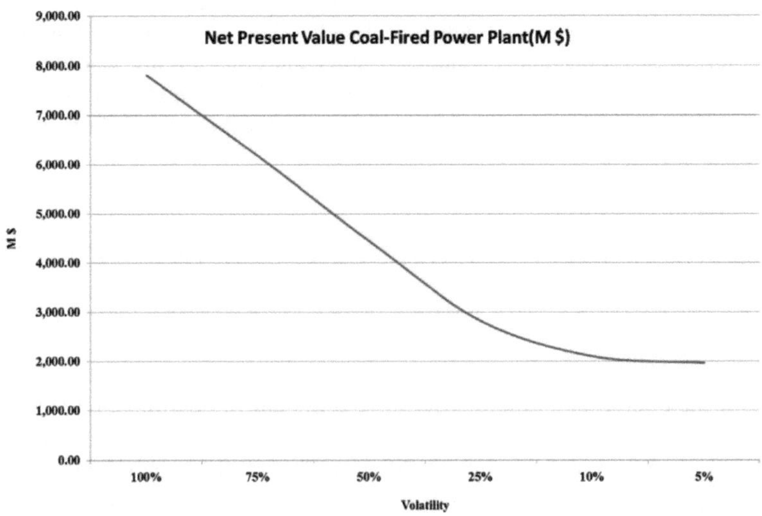

Fig. 11.2 *NPV* of the coal station as function of commodity volatilities

We run 20,000 simulations. Concerning the valuation of the coal plant, Table 11.5 shows the results in both scenarios, i.e. without and with operation flexibility. The flexible operation brings about a positive present value, 7,806.37 M$. This is accomplished by ceasing operations whenever the clean dark spread becomes negative. The value of flexibility amounts to 7,806.37 − 1,388.60 = 6,417.77 M$.

Next we assess how the value of flexibility is affected by the allowance price volatility. Table 11.6 displays the results. As emission allowance prices become more volatile the value of the plant (under flexible operation) grows higher.

Now we consider the case in which the three base volatilities are scaled down in the same proportion; see Table 11.7. Decreasing volatilities leads to lower plant values; they approach the analytical solution for the base load plant (without operating flexibility), $NPV = 1,388.60$ M$. See also Fig. 11.2.

However, in the limit, under null volatilities, the *NPV* would not reach that precise value. The reason is that the expected spread in the far future is negative:

$$\left(E_0(D^E_{54}) - \frac{E_0(S^C_{54})}{0.3877 \times 6.978} - \frac{0.34056E_0(S^A_{54})}{0.3877}\right)e^{-rj\Delta t} - cv^C = -16.08. \quad (11.15)$$

therefore, even without volatility it would be better not to operate. For example, in the final step with $t = 54$ years, $E_0(D^E_{54}) = 146.67$, $E_0(S^C_{54}) = 140.00$, and $E_0(S^A_{54}) = 148.80$. The expected present value of the clean dark spread reached and the investment would not be undertaken. Investing becomes profitable when volatility levels surpass around 25 % of those in the base case.

References

de Neufville R, Scholtes S (2011) Flexibility in engineering design. The MIT Press, Cambridge, MA

Tseng C-L, Lin KY (2007) A framework using two-factor price lattices for generation asset valuation. Oper Res 55(2):234–251

Index

L.M. Abadie and J.M. Chamorro, *Investment in Energy Assets Under Uncertainty*,
Lecture Notes in Energy 21, DOI: 10.1007/978-1-4471-5592-8,

MIX
Papier aus verantwortungsvollen Quellen
Paper from responsible sources
FSC® C105338

If you have any concerns about our products,
you can contact us on
ProductSafety@springernature.com

In case Publisher is established outside the EU,
the EU authorized representative is:
Springer Nature Customer Service Center GmbH
Europaplatz 3, 69115 Heidelberg, Germany

Printed by Libri Plureos GmbH
in Hamburg, Germany